WITHDRAWAL

Play, Creativity and Digital Cultures

Routledge Research in Education

1. **Learning Communities in Education**
Edited by John Retallick, Barry Cocklin and Kennece Coombe

2. **Teachers and the State**
International Perspectives
Mike Bottery and Nigel Wright

3. **Education and Psychology in Interaction**
Working with Uncertainty in Inter-Connected Fields
Brahm Norwich

4. **Education, Social Justice and Inter-Agency Working**
Joined up or Fractured Policy?
Sheila Riddell and Lyn Tett

5. **Markets for Schooling**
An Economic Analysis
Nick Adnett and Peter Davies

6. **The Future of Physical Education**
Building a New Pedagogy
Edited by Anthony Laker

7. **Migration, Education and Change**
Edited by Sigrid Luchtenberg

8. **Manufacturing Citizenship**
Education and Nationalism in Europe, South Asia and China
Edited by Véronique Bénéï

9. **Spatial Theories of Education**
Policy and Geography Matters
Edited by Kalervo N. Gulson and Colin Symes

10. **Balancing Dilemmas in Assessment and Learning in Contemporary Education**
Edited by Anton Havnes and Liz McDowell

11. **Policy Discourses, Gender, and Education**
Constructing Women's Status
Elizabeth J. Allan

12. **Improving Teacher Education through Action Research**
Edited by Ming-Fai Hui and David L. Grossman

13. **The Politics of Structural Education Reform**
Keith A. Nitta

14. **Political Approaches to Educational Administration and Leadership**
Edited by Eugenie A. Samier with Adam G. Stanley

15. **Structure and Agency in the Neoliberal University**
Edited by Joyce E. Canaan and Wesley Shumar

16. Postmodern Picturebooks
Play, Parody, and Self-Referentiality
Edited by Lawrence R. Sipe and
Sylvia Pantaleo

**17. Play, Creativity and
Digital Cultures**
Edited by Rebekah Willett,
Muriel Robinson and Jackie Marsh

Play, Creativity and Digital Cultures

Edited by Rebekah Willett, Muriel Robinson and Jackie Marsh

Routledge
Taylor & Francis Group
New York   London

First published 2009
by Routledge
270 Madison Ave, New York, NY 10016

Simultaneously published in the UK
by Routledge
2 Park Square, Milton Park, Abingdon, Oxon OX14 4RN

Routledge is an imprint of the Taylor & Francis Group, an informa business

© 2009 Taylor & Francis

Typeset in Sabon by IBT Global.
Printed and bound in the United States of America on acid-free paper by IBT Global.

Library of Congress Cataloging in Publication Data

Play, creativity, and digital cultures / edited by Rebekah Willett, Muriel Robinson and
    Jackie Marsh. — 1st ed.
        p. cm. — (Routledge research in education ; 17)
    Includes bibliographical references and index.
    ISBN 978-0-415-96311-4
    1. Play.   2. Creative ability.   3. Creative ability—Social aspects.   4. Digital communication.   I. Willett, Rebekah.   II. Robinson, Muriel.   III. Marsh, Jackie.
    LB1137.P5535 2009
    370.15'7—dc22
                                                            2008018560

ISBN10: 0–415–96311–7 (hbk)
ISBN10: 0–203–88869–3 (ebk)

ISBN13: 978–0–415–96311–4 (hbk)
ISBN13: 978–0–203–88869–8 (ebk)

Contents

List of Tables and Figures   ix

1   Introduction: Encountering Play and Creativity in
    Everyday Life   1
    REBEKAH WILLETT AND MURIEL ROBINSON

PART I
Contexts Of Digital Cultures

Introduction to Part I   13

2   Games Within Games: Convergence and Critical Literacy   15
    CATHERINE BEAVIS

3   Achieving a Global Reach on Children's Cultural Markets:
    Managing the Stakes of Inter-Textuality in Digital Cultures   36
    VALÉRIE-INÉS DE LA VILLE AND LAURENT DURUP

4   Consumption, Production and Online Identities:
    Amateur Spoofs on YouTube   54
    REBEKAH WILLETT

PART II
Children And Digital Cultures

Introduction to Part II   71

5   The Texts of Me and the Texts of Us: Improvisation and
    Polished Performance in Social Networking Sites   73
    CLARE DOWDALL

6   Exciting Yet Safe: The Appeal of Thick Play and Big Worlds   92
    MARGARET MACKEY

7   Online Connections, Collaborations, Chronicles and Crossings   108
    JULIA DAVIES

8   Mimesis and the Spatial Economy of Children's Play Across
    Digital Divides: What Consequences for Creativity and Agency?   125
    BETH CROSS

PART III
Play, Creativity And Digital Learning

    Introduction to Part III   145

9   Creativity: Exploring the Rhetorics and the Realities   147
    SHAKUNTALA BANAJI

10 What Education Has to Teach Us About Games and
    Game Play   166
    CAROLINE PELLETIER

11 Digital Cultures, Play, Creativity: Trapped Underground.jpg   183
    VICTORIA CARRINGTON

12 Productive Pedagogies: Play, Creativity and Digital Cultures
    in the Classroom   200
    JACKIE MARSH

13 Conclusion   219
    MURIEL ROBINSON AND REBEKAH WILLETT

Contributors   229
Index   233

Tables and Figures

TABLES

2.1	Representative Sample of Responses to the Advertisement	27
8.1	Spheres of Game Play	135
11.1	Key Skills and Practices of Participatory Culture	191
11.2	Key Textual Skills and Practices for Participatory Culture	195

FIGURES

3.1	*TS* © Marathon, Mystery Animation Inc.	42
3.2	*Code Lyoko*™ © Moonscoop—France 3-2007—All Rights Reserved.	44
5.1	Clare's profile page, August 2007.	81
5.2	Clare's 'About Me' blurb, August 2007.	81
5.3	Clare's profile page, February 2007.	82
5.4	Clare's 'About Me' blurb, February 2007.	83
5.5	Comments from Clare's profile picture, August 2007.	84
5.6	A page from my mother's photograph album, 1953.	85
8.1	Play diagram of grounds surrounding apartments.	129
8.2	The relations between mimesis and the rhetorics of play and creativity.	140
9.1	3-year-olds making shapes with modeling dough.	156

11.1	Trapped underground	184
12.1	Internet use.	205
12.2	Blog post 1.	206
12.3	Blog post 2.	206
12.4	Blog post 3.	210
12.5	Blog post 4.	212
12.6	Blog post 5.	213
12.7	Blog post 6.	214
12.8	Blog post 7.	214
12.9	Still picture from 'DinoWars 2'.	215

1 Introduction

Encountering Play and Creativity in Everyday Life

Rebekah Willett and Muriel Robinson

The title for this book, *Play, Creativity and Digital Cultures,* gives plenty of scope for exploration. In terms of play, this book includes an examination of video game play, word play on blogs, play with images on video- and photo-sharing sites and play within the context of commercial cultures. Similarly with creativity, the chapters in the book consider a wide variety of circumstances: creative expression through blogs and digital images; creative meaning making across different narrative texts; and creative identity performance through social networking sites. The chapters in this book address some of these topics, but also explore how the terms 'play' and 'creativity' are used in relation to digital cultures.

Theories of play and its relationship with learning are central to our understanding of young children's development (Sutton-Smith, 1997; Vygotsky, 1968, 1978). Winnicott (1971) talked of play as that 'third area in which we are ourselves' which children inhabit between their inner and outer realities. From such research came a new emphasis on the role of play in the education of young children. The drama-in-education movement, spearheaded by practitioner-researchers (Heathcote, 1980), realised the potential for more formalised dramatic role play in children's learning. However, such work has been relatively context-free in terms of its relationship to the new media technologies of modern living (e.g., Livingstone, 2002). Changes in children's media worlds are pointing to a need for a more contextually-located exploration of play as a space which calls upon different digitised global and local landscapes and mixes them in new hybrid forms (Appadurai, 1996).

With regard to literacy, although there have been significant insights into the relationship between play, narrative and reading (Meek, 1991), this work has tended to relate to print literacy rather than using such insights to consider literacy in its wider context. Yet when Meek argues that reading is a form of deep play and that through such play children learn about the power of narrative and of text, there are significant and unexplored resonances with the work being done with regard to play involving new media technologies and the place of popular culture in children's lives. Furthermore, there are connections between literacy research and research in the field of media studies which examines the concept of 'reading' and 'writing' new media texts including computer games and websites.

Over the past two decades we have become increasingly aware of the complex ways in which digital cultures impact on literacy development and on literacy practices outside school. From research in the field of media and cultural studies, we know more about how children make sense of media texts and cultures (Buckingham, 1996; Burn, 2005; Seiter, 2005). Other researchers have added significantly to what we understand about children's media culture as it is (or is not) integrated into the world of school (Dyson, 2003; Lankshear & Knobel, 2003; Marsh & Millard, 2000). Research on new forms of media such as computer games, mobile phones and Internet use is revealing a variety of textualities with which children engage as they play and communicate with peers (e.g., Merchant, 2004; Ito & Daisuke, 2003). Research is demonstrating that creative forms of interaction with media are evident both in children's play (e.g., when combining characters from different media narratives), and through more visible creative productions such as weblogs and home movies (Bromley, 2000; Davies, this volume; Smith, 2005; Willett, this volume).

It is important to acknowledge that a continuing strand of research seeks to demonstrate the deleterious effect of new media technologies on children (e.g., Healy, 1999; Palmer, 2007). Increasingly, others point out that this position starts from a deficit model which sets technologies in competition with each other and sees playful engagement with new media technologies as a distraction, stealing time from the more serious business of engaging with traditional print texts. Instead, the argument goes, we should start from an asset model (Mackey, 2002; Tyner, 1998) which sees texts as essentially porous (Sipe, 2000) and interrelated and all encounters with texts in any medium as allowing children to develop a range of potentially transferable competences.

From the interconnections between these different groups we can see that we need to know much more about the ways in which children play with media texts, how the nature of play itself might be changing, the ways play is repurposing and embedding technology in meaningful everyday social practices and the ways in which new forms of play and creativity relate to classroom practices. This volume addresses questions raised by existing research by presenting and juxtaposing the work of researchers from different disciplines within the international research community and assessing the significance of these ideas. The chapters offer recent insights from research which developed as part of a series of seminars funded by the United Kingdom's Economic and Social Research Council.[1] The seminar series set out to explore the relationships between play, creativity, learning and digital cultures and thereby address the limitations of existing work. As the series progressed, certain themes emerged as particularly compelling and informative, and it is those that are presented in this volume.

Before we offer a framework to contextualise the various contributions here, however, we want to crystallise the tensions and contradictions in the debates by offering some snapshots from our own everyday lived experiences. The vignettes which follow, and our unpacking of their significance, may help readers new to the field to recognise the contested nature of play and creativity as concepts and to see how both concepts are nuanced by encounters with digital cultures.

VIGNETTE 1

> Two 13-year-old girls, Sasha and Aimee, are sitting next to each other on separate computers, interacting in the real space of a cybercafe and also online in a chatroom. One of the common interactions by these girls was 'chasing boys'. Similar to playground games such as 'kiss-chase' or 'catch-and-kiss', they would follow a 'boy' from room to room, trying to get him to talk, particularly trying to get in a 'private conversation'. When chasing a boy, the girls would squeal with delight ('I'm with him, I am, I am, I'm with him'); they would talk about the boys' looks (or more accurately, the look of their avatars); and they would tell each other to stay away, as Sasha said to Aimee (screen name 'aghectic'): 'you lot need to go, this is my man, I'm already chattin to him, do you understand what I'm sayin? Uh uh aghectic you need to move yourself'. The father of one of these girls was particularly enthusiastic about the skills his daughter was developing in the cybercafe. From his perspective, she was typing faster, she was communicating effectively, she could navigate the Internet quickly and easily and she understood the structures of websites.

It is striking, in these particular interactions, how differently this digital play might be constructed. The girls construct their activities as a bit of fun, as indicated in the performative statement Sasha made to Aimee and in statements they made whilst playing, such as 'it's fun to fight over a man'. Clearly they are not in chatrooms to develop their typing skills. From the father's perspective, however, digital play is about developing skills. Here play online is being framed as preparatory, as growth-orientated; that is, play is about making progress, it is about learning. The emphasis is on the social and cognitive skills that are developing. From another perspective, the girls' activities might be considered sexually-charged and therefore risky, and their digital play might be seen as another example in which media is forcing girls to 'grow older quicker'.

From the perspective as a researcher interested in identity, the girls are learning discourses around sexual preferences, gender and age. Barrie Thorne (1993) describes 'kiss-chase' play as a way of setting boundaries between boys and girls, of defining who is a girl/boy and what it means to be a girl or boy. The girls in chatrooms carve out a particular way of 'doing girl', and more specifically doing 'pre-adolescent girl', not only through flirtatious behaviour but also through a way of talking, expressing their opinions and to some extent establishing a particular power relationship with boys. The girls' talk includes statements about how to evaluate boys ('he's really thick', 'eww he's nasty find someone cuter', 'he's a boring old man'). This way of insulting boys (however anonymously), does several things: It gives the girls a position which separates them from the boys, it gives the girls a way of talking about boys and it puts them in the powerful position of being able to decide whether or not to interact with boys.

Coming from this particular perspective, research considers how young people are making meaning from the symbolic resources available to them, how they are negotiating the cultural, economic and political structures surrounding them, how digital play works as a way of performing and defining identities. The focus here is on cultural activity. This is not to deny that young people's involvement in digital cultures involves learning many different skills. However, a view of children and young people's interactions with technology purely in developmental terms, as the girl's father has done, overlooks aspects of digital cultures which may be as important and meaningful as the skills that the young people are developing. By examining how the term 'play' is being used, what narratives it is constructing about technology, and the discursive frameworks that are being drawn on, we are also examining what is being assumed, excluded and what we might be missing as researchers and teachers when we use these concepts to discuss digital cultures.

In his influential book about play, Sutton-Smith (1997) points out that play is always explicitly not the real thing, and therefore children will mark their play with statements such as: 'I'm only pretending'. It is not clear how this pretend space is training for reality—we assume play is connected to that which it is imitating, and that there will be a direct translation into nonplay activities. However, it is not clear that learning how to chase boys in chatrooms, for example, translates into anything other than learning to how to chase boys in chatrooms. In the vignette above, the father's interpretation of improved typing skills might only translate to improved typing within chatroom-like settings. Similarly, the interpretation of play with power relations may not transfer to other areas of the girls' lives. Finally, seeing play as 'not the real thing' calls into question the interpretation of the girls' play as translating into sexualised behaviour offline.

Instead of seeing play as children's work, as preparation for something in 'real life', we might consider ways in which play provides intrinsic consequences such as a sense of well-being or pleasure. Play can be about self-gratification and self-fulfilment. Finally, it is worth considering how constructing play as progress gives adults power to define children's play in particular ways. In Sutton-Smith's words, 'the theories provide rationalization for the adult control of children's play: to stimulate it, negate it, exclude it, or encourage limited forms of it' (1997, p. 49). This is a danger of framing social networking, or any digital activity, in terms of the rhetoric of progress—adult control usurps children's and young people's authority and autonomy whilst at the same time defining what counts as play and ignoring or dismissing all other aspects of play.

VIGNETTE 2

> During one seminar, we had the serendipitous opportunity to see a piece of devised theatre by second-year, undergraduate drama students, called Once Upon A Time. This very clever piece took as a starting point the idea that the characters within a child's fairy tale book were imprisoned in the book, waiting to reenact their tale when the child read the book. We met the characters from Hansel and Gretel and saw their struggle to come to terms with this kind of existence. However, the retelling is interrupted when the child turns from the book to TV, and the characters are possessed by the characters and situations the child encounters as she surfs the channels, finally collapsing back into a catatonic state as they realise their story is not to be completed and resolved. The way in which Hansel, Gretel, their father and stepmother explain this to themselves, is that the child has lost her imagination— television has taken away the possibility of imagining and creating the fairy tale world.

This piece of devised theatre brings together play and creativity in a context where new media technologies and the students' own comfort with a digital culture have been an enabling factor in the success of the piece. The central child character is seen as inherently playful and creativity is a central theme; it is the students' exposure to a wide range of media texts and their own production capabilities using the strategies they have developed as young people growing up in a digital age that enable them to identify a contrasting range of media texts, to sample these electronically and to links these samples into a soundtrack that make the piece feasible.

Once Upon a Time was an assessed task for a third-year module on a BA honours degree called Drama in the Community, which places much emphasis on the notion that the arts are important tools for personal empowerment and for the integration of socially-excluded individuals. As devised theatre, the students are encouraged to see themselves as all capable of devising or writing drama and to draw on their own experiences to do so. Yet the central message of *Once Upon a Time* seems at first sight to be a long way from these underlying principles. How is it that this final year assessed piece from successful students who have spent nearly three years on such a course reverts to a view of imagination as something at odds with experiencing a wide range of texts in different media?

The students responsible for the piece seem simultaneously to hold a range of understandings of creativity, although they do not articulate these and only one is presented explicitly in the script—that of imagination and creativity as stifled by encounters with popular culture. First, from their studies they are aware of creativity as something which is possible for a wide range of people; they themselves have been encouraged to believe that they can produce drama—they see themselves as having the power to be creative, and *Once Upon A Time* demonstrates the truth of this belief. Not only that, their programme sets expectations that drama has a therapeutic and beneficial role to play for disadvantaged groups in society, and they speak at other times of the powerful experiences they have when offering such disadvantaged groups the opportunity to create their own work with no sense that this is in any way a compromised model of creativity. Beyond this tension there is also a sense that for these students the link between play and creativity is also strong—the whole piece is very playful in the sense that it plays with the conventional story and pushes boundaries—so there is another understanding of creativity as inseparable on occasion from play. We can also discern in their work to devise the piece a view of play as a form of modelling of later creative thinking and problem solving in that the plot structure of *Once Upon A Time* requires them to take existing texts and regroup them to make a new set of readings possible. There is of course an irony in that it is their own encounters with a wide range of popular culture texts, reinterpreted in this creative way, which enables the students to make their explicit point about the disabling nature of such texts on a child's imagination—so they are at least tacitly drawing on a more inclusive and empowering model of creativity as possible in popular texts whilst simultaneously making a contradictory statement within the script without any conscious irony.

These two vignettes are of course just that—snapshots of data, presented to illustrate our major themes within the book. But they do prepare the ground for the richly detailed and thoughtful studies which are to follow. Below we summarise in traditional style the territory covered by each contributor; each of the three sections of the book also has a slightly longer section introduction, making connections across the section and between contributors to which we shall return in the conclusion.

INTRODUCTION TO THE CHAPTERS

The aim of this volume is to look at how debates, discussions and research on digital cultures are being framed, particularly in relation to notions of play and creativity. Our starting point is that the concepts of 'play' and 'creativity' are always framed in particular ways—they are never value-neutral. In relation to the anecdotes above, in the first anecdote the girls' play in chatrooms can be defined as 'just a bit of fun' (by the girls), a way of learning ICT skills (by the father) or a meaningful cultural experience (by a researcher). In the second anecdote the students are working with half-understood received views about the dangers of television which would contrast markedly both with their lived experience and on the way they were able to use their own TV experience to transform the piece. These almost chance encounters with everyday views of play and creativity reveal some of the tensions and difficulties we have been exploring through the seminar series, upon which this volume is based. With such fuzzy and contradictory ideas about play and creativity so abundant in everyday life, how are we to make sense of play and creativity, their relationship with each other and our understanding of these concepts in the context of digital cultures?

Because we are seeing children and young people's digital play as cultural activities, we want to suggest that one way the chapters in this book can be understood is through Du Gay et al.'s (1997) model of the 'circuit of culture'. This model theorises how meaning is constructed around a particular cultural object, event or phenomenon through various sites and practices. In the circuit of culture, there are five different interconnected cultural processes or 'moments' where meaning is created and negotiated: production, consumption, regulation, representation and identity. Very briefly, *production* constitutes the economic processes and practices surrounding the creation of a text, including the meanings producers inscribe on texts; *consumption* describes the active process of decoding meaning from texts, involving contexts and negotiation; *regulation* refers to control over activities, ranging from government policies and regulations through to cultural policies and politics (norms and expectations of what is acceptable); *representation* is concerned with signifying practices surrounding a text and the meanings produced through those practices; and *identity* refers to the meanings of a cultural text that develop within a social network.

The moments all interlink with each other in what Du Gay et al. call 'articulations', connections between two or more processes that happen under specific conditions. It is through these articulations that meanings of texts are negotiated and reconstructed. Furthermore, the process of meaning making is ongoing, and therefore meanings are subject to change. Using the Sony Walkman as an example, Du Gay et al. demonstrate that rather than being a piece of technology with a stable and concrete meaning, the Sony Walkman is a site of negotiation, where different practices interact to produce and create ongoing changes in meanings. It is this process of constructing and reconstructing of meaning around particular cultural activities which form the focus of the chapters of this book. The chapters analyse cultural texts or phenomenon that take place at different interlinked moments on the circuit of culture, with each chapter situated within a specific context. We have grouped the chapters together into three sections in a way which highlights the different processes in the circuit of culture. Together, the chapters provide us with a detailed view of how meaning around play, creativity and digital culture is being constructed and negotiated.

Part I, *Commodified Contexts of Digital Culture*, analyses processes of production, looking specifically at the links between production, representation and consumption. This section raises questions about ways commodified contexts of digital culture impact on young people's engagement with digital texts and new media technologies. The focus is on the roles industry and consumer culture play in shaping the ways in which texts are used and interpreted. Catherine Beavis opens this section with her exploration of potential convergences between advertising, youth culture and computer games, and the co-option of ideologies within and outside the game to create new marketing opportunities. Valérie-Inès de la Ville and Laurent Durup analyse how French flavour is given to animated commercial texts through the construction of narratives, plot construction, graphics and artistic choices. Rebekah Willett concludes the section with an examination of how structures from commercial texts are used by young people as they define and perform their identities through videos they produce and post on YouTube.

Part II, *Children and Digital Cultures*, analyses processes of consumption, with links to representation and identity. It contains detailed ethnographic studies of children's engagement with digital cultures, covering a variety of media and contexts. Clare Dowdall opens the section with an analysis of the way one child performs and negotiates her social identities as she works and plays in a range of digital contexts. Margaret Mackey then explores some of the interpretive implications of moving among different media versions of a story, using the concepts of 'thick play' and 'big worlds' as organizers to explore how learning takes place in a world where many stories never seem to come to an end. Julia Davies draws from ethnographic online research spanning across a number of years to describe some of the ways in which individuals use the Internet to play out a range of identities, to experiment with ideas and to make connections with others. Beth Cross concludes the section with an analysis of children's play from on-screen participation to off-screen recombination and transposition of digital motifs, scenarios and tropes in informal role playing activity.

The final part, *Play, Creativity and Digital Learning*, focuses on representations of play and creativity, asking broad questions about how our society views learning and how digital cultures are transforming our views of play, creativity, literacy and citizenship. Shakuntala Banaji opens this section with an examination of the 'rhetorics of creativity', exploring a number of contemporary and persistent political and philosophical traditions in the theorising of creativity, asking whether these claims reflect actual events, trends and practices and whose interests some of these conceptualisations serve. Caroline Pelletier then examines the forms of meaning which a group of young people produced in planning the design of their own video game in an after-school club. Her focus is on conceptions of play, and more particularly the forms of representation and communication that play made possible, highlighting the social function that play performs in negotiating social relations. Victoria Carrington moves us out of schools, but considers implications for education by using examples drawn from a range of digital technologies. Her chapter explores two interesting social phenomena: the emergence of new forms of civic participation; and, shifts in our view of play and creativity. She asks what these examples tell us about the influence of digital technologies on the everyday life of our society and as a consequence, their significance for school-based literacy practices. The section concludes as Jackie Marsh explores the potential relationship between digital literacy and play within a 'productive pedagogies' (Lingard, Ladwig, Mills, Bahr, Chant et al., 2001) model, demonstrating how, through the facilitation of playful approaches to a digital literacy curriculum, educators can develop a pedagogical approach that ensures learner agency, curriculum relevance and intellectual challenge.

The key aims of the book throughout these sections will be to address and provide evidence for continuing debates around the following questions:

- What notions of creativity are useful in our fields?
- How does an understanding of play inform analysis of children's engagement with digital cultures?
- How might school practice take account of out-of-school learning in relation to digital cultures?

In our conclusion we shall return to these questions and offer areas for further debate and research.

NOTES

1. The seminar series entitled, 'Play, Creativity and Digital Cultures', was funded by the UK's Economic and Social Research Council, Award reference: RES-451-25-4151, and ran from 2005–2007. The series website is available at http://www.bishopg.ac.uk/?_id=865&page=7

REFERENCES

Appadurai, A. J. (1996). *Modernity at large: Cultural dimensions of globalization*. Minneapolis, MN: University of Minnesota Press.
Bromley, H. (2000). Never be without a Beano! Comics, children and literacy. In H. Anderson & M. Styles (Eds.), *Teaching through texts* (pp. 29–42). London: Routledge.
Buckingham, D. (1996). *Moving images: Understanding children's emotional responses to television*. Manchester, UK: Manchester University Press.
Burn, A. (2005). Potter-Literacy—from book to game and back again; literature, film, game and cross-media literacy. *Papers: Explorations into Children's Literature, 14*(3), 5–17.
Du Gay, P., Hall, S. Janes, L. Negus, K., MacKay, H. and James, L. (1997) *Doing Cultural Studies: The Story of the Sony Walkman*. London, UK: Sage.

Dyson, A. H. (2003). *The brothers and sisters learn to write: Popular literacies in childhood and school cultures.* London: Teachers College Press.

Healy, J. (1999). *Endangered minds: Why children don't think and what we can do about it.* New York: Pocket Books.

Heathcote, D. (1980). *Drama as context.* Sheffield: National Association for Teaching English.

Ito, M., & Daisuke, O. (2003, June 22–24). Mobile phones, Japanese youth, and the re-placement of social contact. Paper presented at Front Stage–Back Stage: Mobile Communication and the Renegotiation of the Social Sphere Conference, Grimstad, Norway. Retrieved October 17, 2007, from http://www.itofisher.com/PEOPLE/mito/mobileyouth.pdf

Lankshear, C., & Knobel, M. (2003). *New literacies: Changing knowledge and classroom learning.* Buckingham, UK: Open UP.

Lingard, B., Ladwig, J., Mills, M., Bahr, M., Chant, D., Warry, M., et al. (2001). *The Queensland School Reform Longitudinal Study* (Vols. 1 and 2). Brisbane: Education Queensland.

Livingstone, S. (2002). *Young people and new media: Childhood and the changing media environment.* London: SAGE.

Mackey, M. (2002). An asset model of new literacies: A conceptual and strategic approach to change. In R. Hammett & B. Barrell (Eds.), *Digital expressions: Media literacy and English language arts* (pp. 199–217). Calgary, Canada: Detselig Enterprises.

Marsh, J., & Millard, E. (2000). *Literacy and popular culture in the classroom.* London: Paul Chapman.

Meek, M. (1991). *On being literate.* London: The Bodley Head.

Merchant, G. (2004). Imagine all that stuff really happening: Narrative and identity in children's on-screen writing. *E-learning, 3*(1), 341–357.

Palmer, S. (2007). *Toxic childhood: How the modern world is damaging our children and what we can do about it.* London: Orion Publishing Group.

Seiter, E. (2005). *The Internet playground: Children's access, entertainment and mis-education.* New York: Peter Lang.

Sipe, L. R. (2000). Those two gingerbread boys could be brothers: How children use intertextual connections during storybook readalouds. *Children's Literature in Education, 31*(2), 73–90.

Smith, C. R. (2005). The CD-ROM game: A toddler engaged in computer-based dramatic play. In J. Marsh, (Ed.), *Popular culture, new media and digital literacy in early childhood* (pp. 108–125). London: RoutledgeFalmer.

Sutton-Smith, B. (1997). *The ambiguity of play.* London: Harvard UP.

Thorne, B. (1993). *Gender play: Girls and boys in school.* Buckingham, UK: Open University Press.

Tyner, K. (1998). *Literacy in a digital world: Teaching and learning in the age of information.* Mahwah, NJ: Lawrence Erlbaum Associates.

Vygotsky, L. (1968). *Thought and language.* Cambridge, MA: MIT Press.

Vygotsky, L. (1978). *Mind in society: The development of higher psychological processes.* (Michael Cole, Ed.). London: Harvard University Press.
    The whole book is an edited translation from the Russian from a series of essays by Vygotsky. The editors are actually Michael Cole, Vera John-Steiner, Sylvia Scribner and Ellen Souberman. There is a one-page editor's preface which explains the provenance and identifies the various translators and a 15 page introduction by Cole and Scribner. The rest of the book is by Vygotsky.

Winnicott, D. W. (1971). *Playing and reality.* London: Tavistock Publications.

Part I
Contexts of Digital Cultures

Introduction to Part I

The aim of this section is to consider contexts of children's play and creative engagement with digital cultures which are sometimes left out when looking at specific play situations. In particular, this section focuses on industries, marketing and young people's understanding of and engagement with structures and strategies from commercial texts. These chapters go beyond a polarised view of children as either dupes to manipulative marketing strategies or as powerful and fickle consumers. Instead the chapters examine the complex convergences that are occurring within the texts and products themselves, and also within audience's engagements with commercial texts.

The section opens with a chapter by Catherine Beavis which examines young people's understandings of new convergences between digital media cultures and marketing strategies. This chapter explores potential convergences between advertising, youth culture and computer games, and the co-option of ideologies within and outside games to create new marketing opportunities. It takes the example of advertising that merges multiplayer games, girl power and soft drink marketing to raise questions about how young people read and respond to texts of this kind, and how teachers might work with them in ways that acknowledge their power and pleasures, while also developing awareness, understanding and critique.

Next, the chapter by Valérie-Inés de la Ville and Laurent Durup analyses market strategies which rely on different types of inter-textual references. The chapter analyses two French animations, examining the interplay with Japanese standards and American archetypal narratives that are recognised by international audiences. Here questions about global marketing strategies and connections to children's cultural worlds in different countries are explored.

The section concludes with a chapter by Rebekah Willett which examines young people's play with media texts, looking specifically at amateur spoofs or parodies which are distributed online. The focus of the analysis in this chapter is on how the structures from commercial texts are used by young people as they define and perform their identities. The chapter engages with theories about play and convergence culture to analyse young people's online consumption and production activities.

Together these chapters offer a view of children and young people's play with digital texts which is contextualised by particular structures within the texts themselves, structures which are tied to commercial industries and converging with children and young people's cultures.

2 Games Within Games
Convergence and Critical Literacy

Catherine Beavis

For some time now there has been considerable interest in, and awareness of, the centrality of popular culture and new media in many young people's lives and of the need for teachers and researchers to recognise and respond to this in relevant and productive ways (e.g., Buckingham, 1998, 2000; Sefton-Green, 1998; Alvermann, Moon, & Hagood, 1999; Marsh & Millard, 2000). Schools and systems are increasingly coming to recognise the need for curriculum to engage with the 'changed communicational landscape' of the present day (Kress, 2000), and to develop strategies and frameworks to help students and teachers become more critically reflective and analytic about multimodal forms of text and literacy. In Australia, English and literacy curriculum propose an expanded view of literacy that incorporates both print and digital forms. The national *Statements of Learning for English* (Ministerial Council for Education, Employment, Training and Youth Affairs [MCEETYA], 2005), which outline common elements underlying English and literacy curriculum in each Australian State, utilise Luke, Freebody, and Land's definition of the literacy young people need in the contemporary world as 'the sustained and flexible mastery of a repertoire of practices with the texts of traditional and new communications technologies via spoken language, print and multimedia' (Luke, Freebody, & Land, 2000, p. 20; MCEETYA, 2005, p. 4).

In research exploring the texts and literacies of online culture, attention is paid to both texts themselves and to informal learning, and what might be learnt from observing young people's engagement with popular culture out of school. Studies of out of school engagement with digital culture and computer games focus on the textual nature of computer games, the kinds of learning and affiliations young people are involved in as they play computer games, situated learning and practices around game play, representations of self, the construction of communities, relationships and identities in 'real' and 'virtual' worlds, and the ways in which game playing functions as an arena for the development and display of competence (e.g., Facer & Williamson, 2004; Gee, 2003; Schaffer, Squire, Halverson, & Gee, 2005; Steinkuehler, 2004). These studies point to understandings and expectations about texts, participation and community that position young people as agential and knowledgeable in the digital world, but in areas that for the most part go unrecognised and unaddressed by schools.

At the same time, the ways in which students are themselves shaped and positioned within the globalised world of media culture means that pedagogies designed to prompt students to develop critical reflective analysis of popular culture and new media need to take account of the intricate intermeshings between these texts and technologies and the ways people live their lives. As Nixon (2003, p. 407) describes it, 'participation in global popular media culture, including online culture, has become integrally bound up with children's and teenagers affiliations, identities and pleasures' with 'this kind of social participation . . . integrally bound up with the ways in which symbolic meanings are made, negotiated and contested'.

The development of critical literacy is a central component of much English curriculum. Its goal is to help students recognise and deconstruct both the structural features of the text and their interaction, and the way the text works to construct a particular set of values or ideology and persuade the reader/viewer to share those views. If curriculum is to address the kinds of texts and literacies young people engage with in their everyday worlds, popular culture provides a logical and readily accessible site for doing so. However, texts drawn from popular culture need to be introduced and approached in ways that recognise both the flattening effect of classrooms on any text, and the ways in which transplanting a text into this environment alters the ways in which it connects with other texts and how it is read. Making a popular text the subject of school-based analysis runs the risk of stripping the text of many of those qualities that made it popular in the first place, and of constructing student viewers or readers as naïve, ill-informed or transgressive. This suggests that 'playful pedagogies' (Buckingham, 2003), which acknowledge the role of both pleasure and complicity in young people's engagement with digital culture, and include opportunities for 'irreverent play with meaning in which seriousness and rationality are replaced by irony and parody' (Buckingham, 2003, p. 17), are more likely to be successful with teenage students in particular than teaching based on more traditional, more serious-minded approaches to critical and media literacy.

CONVERGENCE AND MEDIA LITERACY

A core feature of the 'changed communications landscape' is that it is characterised by convergence and continual change. While convergences between video or computer games and other media forms—blockbuster films and book bestsellers such as *Lord of the Rings*, *Harry Potter*, James Bond and so on—are well established, the uses and opportunities of convergence extend well beyond the cross platform adaptation of narrative. Convergence 'facilitates the flow of content across the entire media system'

(Convergence Culture Consortium, 2006, n.p.). Media convergence incorporates ownership, globalisation, marketing, audience experience and the translation of products across media forms. The term:

> refers both to shifts in patterns of media ownership and to the ways in which media is experienced and consumed. Media convergence is more than simply a technological shift. Convergence alters the relationship between existing technologies, industries, markets, genres and audiences. Convergence refers to a process and not an endpoint. . . . Convergence is taking place within the same appliances . . . within the same company . . . within the brain of the consumer . . . within the same fandom. . . . Convergence is more than a corporate branding opportunity; it represents a reconfiguration of media power and a reshaping of media aesthetics and economics. (Jenkins, 2004, pp. 34–35)

This concept of convergence has considerable explanatory power in describing global media culture, and provides a suggestive framework for teaching media literacy. The Consortium develops two frameworks that help map the dimensions they describe. The first outlines intersections between 'transmedia entertainment', 'participatory culture' and 'experiential marketing':

> *Transmedia Entertainment* describes the newfound flow of stories, images, characters, information, and sounds across various media channels, in a coordinated fashion, which facilitates a deepening expansion of the consumer's experience.
>
> *Participatory Culture* describes the way consumers interact with media content, media producers and each other as they explore the resources available to them in the expanded media landscape. Consumers become active participants in shaping the creation, circulation and interpretation of media content. Such experiences deepen the consumer's emotional investment in the media property, and expand their awareness of both content and brand.
>
> *Experiential Marketing* refers to the development of novel approaches to brand extension and marketing which play out across multiple media channels so that the consumer's identification with the product is enhanced and deepened each time they re-encounter the brand in a new context. (Convergence Culture Consortium, 2006, n.p.)

The second framework refers to the interplay between 'three key, yet different, cultural groups: fan cultures, brand cultures and style cultures':

> *Fan cultures:* the social groups of impassioned consumers of specific genres of entertainment (examples include science fiction, comic books and computer games).

Brand cultures: communities of committed consumers of specific products and services (such as BMW drivers, iPod users and Coke collectors).

Style cultures: the neo-tribes that organise around the ephemeral pursuit of cool. (Bobos in China, Hip Hop Hedz in America and Otaku and Kogal in Japan). (Convergence Culture Consortium 2006: n.p.)

The elements and intersections mapped in these two clusters provide a useful framework for identifying ways in which convergence of this kind calls on existing identifications and affiliations to build new markets. As such, it also suggests a framework for critical analysis and 'playful pedagogies' of the kind Buckingham describes.

WORLD OF WARCRAFT MEETS COKE

In the summer of 2005 multinational companies Blizzard and Coca-Cola joined forces to launch the Massively Multiplayer Online Role Play Game, *World of Warcraft*, and the soft drink, Coke, on the Chinese mainland simultaneously. The centrepiece of the campaign was a lavish animation designed for television and also located on the *World of Warcraft* community site and (at the time) on the China iCoke websites (http://www.world-ofwarcraft.com/community/chinaicoke.html; see also http://www.youtube.com/watch?v=Vc8rWbplKhg). It began with a confrontational board meeting between the three young women who form the (real life) highly popular 'Chinese pop sensation' SHE, and an aggressive and unpleasant manager, who tells them 'sexy is what sells. You get my drift?' The young women, united at the opposite end of the table, look up from their laptop where they have been playing *World of Warcraft*. As the women look up, the camera reveals an image of a *World of Warcraft* ogre on their laptop. As one, they reply 'No'. When the manager insists, 'what do you mean no'? they reply 'No means no'. He and they morph into *World of Warcraft* characters, the cans of Coke the girls have been drinking transform to swords, and they do battle in the epic landscapes of the *World of Warcraft* world. After a brief skirmish, set to thunderous music, the three women strike the ogre into oblivion, open a chest full of Coke, and drink silhouetted against the sky. In a flash, we are returned to the 'real' world, where we see the manager, now in his underpants, gratifyingly defeated and abashed.

The advertisement, launch, and related sites and activities, provide a compelling instance of media and cultural convergence. The campaign capitalised on the popularity and 'cool' of both products, global multinational partnerships, cross-platform referencing between internet and television and on and offline forms. It also drew on and sought to extend the allegiances, fan cultures and global networks within and around this immensely popular Massively Multiplayer game. An announcement of the launch, together with photos from it, was posted on the *World of Warcraft* site:

> *World of Warcraft* recently launched with amazing fanfare and success in China. Partnering with *Blizzard* to usher in this blockbuster launch was *iCoke*, Coca-Cola's name-brand presence in China. In the weeks leading up to and after the launch, *iCoke* led the way in promoting *World of Warcraft*, celebrating the game in real-life replicas of Stormwind Castle and Thunder Bluff, creating a joint *World of Warcraft* and *Coca-Cola* televised commercial, featuring Chinese pop sensation *SHE*, and unveiling an *iCoke* branded *World of Warcraft* site. (http://www.worldofwarcraft.com/community/chinaicoke.html)

Photos showed crowds of people queuing up to enter the launch site, mock ups of *Warcraft* locations, people dressed as *Warcraft* characters, snapshots of attendees, multiple images of Coke drinkers and the Coca-Cola logo and a giant poster of a stylised female figure from the game that also appeared on numerous magazine covers in a further layer of blanket media coverage.

AdAge China, a Chinese trade magazine, described the joint marketing campaign:

> China has an estimated 40 million gamers, mostly teens and young adults, in *Coke*'s target market and 'will probably be the world's biggest online gaming market in the world by next year. This is an exploding passion point' said Ilan Sobel, Coca Cola's Shanghai-based general manager, strategic marketing and innovation for China. . . . *Coke* is essentially in the entertainment business, our brand is about bringing a little happiness into the lives of our customers. With the surging popularity of the internet, we hope to further our connection with the new generation through this passion point and to continue to galvanise our relationship with teens. (Madden, 2005, n.p.)

There are a number of ways one might think about this campaign and the phenomenon it represents. The advertisement exemplifies the network of kinds of concepts and relationships the Convergence Culture Consortium describe, with intersections between transmedia entertainment, participatory culture and experiential marketing. The two main narratives, nested within each other, of the girls' refusal of the boss's demands, and the in-game battle and victory, effectively call on existing allegiances and investments, while also promising participation in new and exciting communities. In particular, the interlinking of fan narratives around the pop group SHE and World of *Warcraft* with the brand culture of Coke and *World of Warcraft*, and the promise of both to create 'cool' and contemporary hip global identities, is heavily emphasised in both the advertisement and the website.

Computer Mediated Communications, and particularly Massively Multiplayer Online Roleplay Games (MMORPGs) such as *World of Warcraft*, appear to generate high levels of emotional investment, complex social interactions and strong social bonds and relationships (Turkle, 1995; Yee, 2005, 2006). The advertisement calls on intersections of fan, brand and style cultures much as the Convergence Culture Consortium describes. Its use of *World of Warcraft* potentially taps into the vast numbers and allegiances of *World of Warcraft* players worldwide to support market growth for the game's presence in China, as well as for Coca-Cola. According to Yee, young players of Massively Multiplayer games regard their experiences within the games they play as having considerable significance in their lives:

> Of all age cohorts, users under the age of 18 are most likely to feel that the friendships they have formed online were comparable or better than their real-life friendships, and they were also most likely to self-report that the most positive or negative emotionally salient experience they have had in the last month as having occurred in the MMORPG environment rather than in real life. This is also the cohort that felt they had learned the most about leadership skills from the MMORPG environment. (Yee, 2006, pp. 35–36)

The possibilities offered by this combination of deep immersion in the world of the game, the long hours spent there, the importance of games in players' lives, and the existence of strong social bonds and extensive networks are beginning to be recognised by mainstream marketing. Online business sites urge companies to look to online games as vast and untapped markets—'a promising new marketing medium [that] has crept up behind us . . . [whose] enormous potential as advertising channels has been tapped only in a limited way' (Lindstrom, 2007, n.p.). Advertisers are urged to look not just at the size and demographic Massively Multiplayer games offer, but also at the ways in which the immersive qualities of games offer further leverage. This can be achieved most effectively where companies can be 'creative' and fuse brands with 'attributes', consistent with the kinds of fusion and convergence the Convergence Culture Consortium model describes:

> Marketers looking for 18 to 34 year old males can look to massively multiplayer online games (MMOG) for opportunities, according to a Universal McCann report, 'TrendMarker, Parallel Worlds'. While these opportunities exist in certain genres, brands must be creative to make an impact in certain massively multiplayer online game (MMOG). . . . In social and real-world-based games standard ads add authenticity, though some marketers opt for deeper integration. Electronic Arts'

'Sims Online' offered Nike shoes that enhanced a player's speed. Within 'Second Life,' players can buy MP3 music from iTunes and play the songs in the game. The report sums up this level of engagement with the suggestion 'interaction with the virtual world allows marketers to fuse brands with attributes in a powerful way'. (Burns 2006 n.p.)

'A Forrester Consumer Technographics report says 21% of North American consumers regularly spend leisure time playing games. The MMOG category takes up a significant portion of that, offering a social, engaging experience in which marketers can reach consumers'. (Burns, 2006, n.p.)

The Universal McCann report highlights the 'hyperconnected audience' as both 'opportunity and a threat': 'The high levels of engagement mean that users are attentive, ready to react, and their strong P2P relationships mean that a positive endorsement will be passed around quickly' although 'instances of players staging virtual sit ins and riots to protest at games publishers during disagreements have been known'. Similarly, while 'opportunities to buy the rights to translate real world properties into the virtual environment are relatively expensive', there are 'opportunities to create virtual marketing campaigns in collaboration with the players' (Universal McCann, 2006, p. 7).

The attractiveness of games as advertising markets, and their quickness of business to identify the potential offered by the interaction and 'social engaging experience' games offer, provides powerful arguments for working with students to develop critical awareness of the presence and consequences of advertising within and around the games world.

While the young players of Massively Multiplayer games market are still predominantly male, albeit within increasing numbers of female players, the WOW/Coke advertisement draws on girl power ideologies and the attractive presence of well-known stars to ensure both males and females remain engaged. The SHE characters take no nonsense from exploitative males, stand up for right values and embody the laudatory sentiment of the Coke slogan—'for satisfaction, look to yourself'. As typifies girl power, their behaviour is heavily contradictory. They are both daring and conservative in their defence of autonomy and self-reliance. While they fight to resist the managers' demand that 'sexy is what sells', they do so wearing the stereotypically skimpy and provocative dress that characterises so many female role play game avatars. Even the slogan, at least in this translation, is charged with overtly sexual intimations at the same time as it asserts autonomy. This ambiguity ensures the women can be seen as both sexy and resistant; feminist but also dutiful; contemporary but also affirmative of traditional values, and so on. As Hopkins notes of popular culture's fascination with girl heroes and girl power, 'the 'radical' is a rare commodity and everyone wants a piece of it' (Hopkins, 2002, p. 8).

GLOCALISATION AND HYBRIDITY

As the introduction of both *World of Warcraft* and Coke to mainland China, the advertisement relies on the reputation of both to activate interest and generate new markets, but must also appeal to consumers in very different cultural settings. As de la Ville and Durup argue elsewhere in this volume, successful marketing internationally of cultural products such as animations depends on the suggestive power generated through 'a complex inter-textual elaboration, which allows a creative interplay with the Japanese standards or American archetypal narratives that international audiences recognise' (de la Ville and Durup, this volume). Although pitched at an older audience than the television animation series de la Ville and Durup analyse, *World of Warcraft* in many ways epitomises the effectiveness and marketing power of 'the French touch' they describe. The experiential appeal of the advertisement depends to a large degree on the aesthetics and narrative of the game to carry the right connotations, and to strike the right balance between local and global—utilising narrative, plot construction, graphics and artistic choices in the ways de la Ville and Durup describe. In this, a third and fourth set of convergences, in addition to those suggested by the Convergence Culture Consortium, are evident: the reconstruction of local/global relationships and a version of aesthetics familiar from console and computer games sometimes referred to as Pan-Asian or Faux-Asian (Consalvo, 2006).

The relationship between local and global commerce and cultures is complex. Contrary to initial expectations about global marketing within the advertising industry, that higher income would lead to homogenous consumer needs, tastes and lifestyles (De Mooij, 2001,), this has not proved to be the case. Rather, 'with converging wealth, convergence of consumption turns into divergence' (De Mooij, 2001, p. 188). Similarly expectations—that global teenagers would emerge as a group with stronger links and cultural identities between each other than with their national identities—also appear oversimplified. Convergence and divergence, De Mooij argues, are taking place at macro levels, but to different degrees and in different communities (De Mooij, 2001, p. 189). She cites Coca-Cola as one example of large multinational firms who 'have seen their profits decline because centralised control lacks global sensitivity and are consequently changing their strategies from local to global' (De Mooij, 2001, p. 184).

Marketers such as De Mooij warn against assuming universal values, arguing rather that cultural values remain strong and distinct, and that global marketers ignore these at their peril. Consalvo (2006) by contrast, warns against seeing essentialist or fundamental national qualities, arguing rather that culture 'is mobile rather than static, and has always sought the influence of whatever is new, different, "foreign" or strange. . . . A culture that does not change is a culture that is dying' (Consalvo, 2006, p. 134).

The *World of Warcraft*/Coca-Cola advertisement seems to embody both these points of view. While the young women provide a strong local point of identification, the images themselves, and the products, Coca-Cola and *World of Warcraft,* are more representative of European and American aesthetics, and rely on this for part of their appeal. The ideologies, however, particularly in the English translation that appears as subtitles on the advertisement, seem universal, as in the 'no means no' translation, or the Chinese Coke slogan—'for satisfaction, look to yourself'. Yet both are only rough approximations of their Western equivalents. The 'no means no' resonates as a familiar mantra in Western feminism, asserting women's right to choose the nature and level of their relationships with men. As such, it strikes a chord, signalling that it is highly appropriate to the young women's refusal of the manager's request for a 'sexy' campaign (if not to the forms of dress in which they subsequently do battle), and it acts as a war cry for the righteous battle of self defence on which they embark. It would be useful to know however, the degree to which, for Chinese audiences, the phrase also calls on darker, more specific Western meanings, rebutting unwanted and insistent sexual requests, and whether associations such as these would alter the advertisement's tenor and tone significantly.

The advertisement is thus an interesting exemplar of the kinds of national and cultural hybridity, and the interaction of local and global, that Consalvo (2006) describes with respect to the video game industry, particularly console games. Hybridity in this context, she argues, encompasses two types of 'fusion':

> the melding of business and culture as well as a convergence between Japanese and US interests. . . . Just as different national identities have been mixed in the hybrid, so too the realms of business and culture are converging in novel ways. . . . This hybrid . . . is a significant point at which a global media culture is created that is unlike any national media culture in its composition. (p. 120)

She calls on Robertson (1995) to describe the notion of 'glocalisation': 'the successful global transfer of products to different localities by making modifications for such variables as culture, language, gender or ethnicity . . . [where] the local should not be seen in distinction to the global, but that instead both are mutually constitutive' (Consalvo, 2006, pp. 120–121). Similarly, she cites Sony's strategy of 'global localisation' (Morely & Robins, 1995, p. 150) 'which entails gaining "insider" status within regional and local markets as it operates around the world' (Consalvo, 2006, pp. 120–121).

The pan-Asian or 'faux-Asian' aesthetic she describes as characteristic of Japanese influenced global media culture is present, oddly inverted in this context as a reimportation of Asianised images back into (different parts of) Asia through the Western medium of *World of Warcraft*. In this instance the East/West images and aesthetic in the advertisement have a Western inflection, and serve the purposes of 'glocalisation' within China by collocating these bodies, gestures and faces with the 'coolness' offered by Coke. In addition to the clip itself, the iCoke website juxtaposed further sets of related images mixing 'pan-Asian' and more European or American aesthetics.

By contrast to the WOW/Coke advertisement, which uses settings and characters directly drawn from the game, the web-based *World of Warcraft* set of screens originally posted on the Chinese iCoke site, seemed more strongly Asian than the game itself, in its drop in props and characters and colours and composition and in the women's faces and bodies. The three women, dressed in red, are grouped front centre, wielding a staff, a bow and a sword and wearing thigh boots, armbands and skirts and shorts of various lengths. In the right foreground is the red chest of Coke, with two large red banners with Coke insignia and two lampposts a little behind. On the left, a wooden signpost with Chinese inscriptions shows multiple pathways. The women are in a forest clearing with a pebbled floor; behind them the forest is shown in shades of green, aqua and purple, with leaves ranging through pinks, mauves, dark blue and white, and what looks like a monster in the far left background, eating leaves, while behind it, in turn an intense white light glows. A wooden sign suspended by chains drops down about the left hand signpost—it has the *World of Warcraft* emblem of a W within a circle and Chinese writing underneath. In the upper right hand corner is a more elaborate rendering of the *World of Warcraft* trademark, gold on a sky blue background, with the website underneath. On the far right, a vertical strip with instructions in Chinese, topped by an arrow, points the viewer into subsequent screens. (see http://icoke.sina.com.cn/moshou/index.html)

A second set of screens took viewers to contemporary China and a vision of the excitement offered by Coca-Cola in the contemporary world. Here, a different again set of aesthetic choices and amalgams characterised the group of cool kids that were the 'real life' face of Coke; in many respects extremely Westernised in hair, gesture, dress and so on. Sets of photos and other graphics in red, gold, black and white are set on a white screen, against a larger red background. Further screens show more shots of this group out on the town, bottles, logos, advertising claims and so on. The inter-textuality and cross-referencing at narrative, graphic and iconic levels, evident in the advertisement, the website images, website photos from the launch and the images of young Coke drinkers on the iCoke website, exemplify the successful kinds of cultural and aesthetic hybridisation both de la Ville and Durup (this volume) and Consalvo (2006) describe. The hybridity evident within the advertisement is translated to a further level of naturalisation beyond animation and into the dress, stance and portrayal of actual fans and actors captured on the websites for iCoke and the game.

WORLD OF WARCRAFT AND YEAR EIGHT

The appearance of this advertisement on the *World of Warcraft* site high-lighted questions about whether the location and content of the advertise-ment provided a means for advertisers to bypass conventional frameworks of scepticism or resilience that accompany much response to advertising in the mainstream media, in particular print or television advertisements that are more traditionally the stuff of critical literacy and media analysis in school. If schools are to take seriously extended definitions of literacy, and the need to help students develop skills of critical analysis of multimedia and electronic texts, the *World of Warcraft*/Coke advertisement provided a timely opportunity to explore how students viewed such advertising, how they located it within broader frameworks of in and out of school experi-ences of both advertising and digital texts, and how effective they under-stood such approaches to be.

Early in 2006 I had been in discussions about English and computer games with two teachers at a large Melbourne boys' school. I was keen to learn more about what they were doing with their year eight classes, and about how these 14–15-year-old students thought about the appearance of advertisements within computer games. Research goals for the project were to:

- Learn more about the nature of multimodal (non print) forms of texts and literacies, in line with developments in English curriculum guide-lines in Victoria;
- Investigate the ways in which popular computer-based texts, such as computer games, are used as advertising and marketing tools, singly and in combination;
- Understand more about how young people read and respond to texts of these kinds;
- Explore approaches to teaching and studying such texts in English to help students understand more about how meanings are made and values carried in multimodal popular texts such as computer games; and
- Develop frameworks for use with students to help them recognise and critically analyse texts where different forms combine for advertising purposes.

The research team consisted of myself, a teacher educator at Deakin University, Melbourne, and Claire Charles, research assistant to the proj-ect and a doctoral student at Monash University. The two teachers, James and Tim,[1] agreed to introduce the advertisement to their classes, as part of their 2-week unit on computer games. Prior to this unit, students had done extensive work on narrative, film and advertising. Over a double period, students were shown an initial *World of Warcraft* trailer to contextualise the advertisement. Following whole class discussions of the trailer, exploring aspects of the trailer's appeal and the ways it worked as both narrative and advertisement to entice players to buy the game, students broke into groups for more work around the *World of Warcraft*/Coke advertisement. They were given worksheets and activities, and one group at a time, shown the advertisement on a laptop computer. In a later class, students were interviewed in small groups about their preferences as game players, their responses to the advertisement, and their opinions about the phenomenon of converging advertising and computer games more generally.

At the start of the unit, James had surveyed his class and found that all but one owned some form of gaming console or X-box, that a third of them said they spent more time chatting on MSN or other online social networking sites than talking face to face, and half the class spent more time playing on computers than watching TV. The boys in Tim's class were likewise almost all players of games of some kind. However, the games the boys in both classes played were overwhelmingly sports or car driving games—action oriented and realistic—by preference to more fanciful and 'unrealistic' RPG and MMORPG games. None of the students interviewed played these latter games on a regular basis, and only a handful had heard of *World of Warcraft* previously. However, almost all responded enthusiastically to the *World of Warcraft* trailer at the start of the lesson, and participated actively in discussions about how it worked to attract players to the game.

As a marketing tool for the game itself, with this group the clip worked admirably. In interviews, many boys spoke of now wanting to buy and play *World of Warcraft*, making an exception to their rule of avoiding fantasy-based 'medieval' role-play games. Several joked they wanted to drink Coke and play the game at the same time, from their tone only half self-deprecating in their recognition that this was just what viewers of the clip were 'supposed' to want to do.

As part of their work around the advertisement, we asked the students to fill out a chart showing what they liked and disliked about it. A representative sample of answers (as shown in Table 2.1) shows favourite features included the shift to the Warcraft world, the fighting, the girls, the humiliation of the boss and the concept. Some students actively disliked the advertisement, citing amongst other things the very concept, the fact that it was for an Asian audience, and that it was unrealistic and 'disconnected from anything'.

For all *World of Warcraft*'s popularity, none of the boys in our study at that time played the game, and a number had never heard of it. Their responses rely in part on the introductory trailer shown to them prior to the advertisement, and also on work done in class earlier on advertising, computer games and advertising, as well as their own extensive experience of sports and car racing games. Their responses show a mix of elements. Positive responses include 'lots of action with women involved'. They are

Table 2.1 Representative Sample of Responses to the Advertisement

What do you like about the clip?	*What don't you like?*
What I liked about the clip is the action, the violence and the fighting. It showed that the characters had a reason for fighting.	I don't like the way they assumed that we knew about the game, because if you didn't know anything about the game, you wouldn't understand the ad fully.
The things I like about the clip is there are hot chicks, they drink *Coke*. I also like how they go into battle and beat the boss.	I liked all of it.
I liked the way the producers were able to advertise *coke*, a modern day drink, with *World of Warcraft*, a mythological game with no technology.	I didn't like the way that when they were in the office and trying to sell something, how it was too quick for you to get an idea of what's going on.
I like the bit where they take on the boss and beat the boss up and after they beat him up they open the chest of *Coke* to regain their power.	They should introduce every one of the characters and make it more action packed.
The only thing I 'liked' about the ad was that it was targeted at a specific target market of *World of Warcraft* players. The girls in the ad being transformed into princesses in *World of Warcraft*, was meant for young Asian boys who play Warcraft all day, and the fact that they saw that market is the only thing I like. A very well planned ad for that market. And only that market.	Personally, I hated the ad. I know as a fact that none of my friends would be more inclined to buy *Coke*. It is a signature, quirky, Asian ad like many others.
Nothing.	One thing I don't like about this clip is that there was no and I mean no absolute meaning to it. There was a bit of fighting and a bit of sexism involved in the clip.
The main things I like about the clip are that there is a lot of action with women involved. I liked the *Coke* factor of this ad, and how the 3 women embarrassed the boss with his pants around his ankles.	I did not like the ease of beating the big ogre and how everything just went into this imaginary world without purpose or reason.
I like the way they just break out into a war and become these cool monsters of *WOW* [*World of Warcraft*]. I also like how they are rewarded with it.	I didn't like the fact that they didn't do a sexy ad.

interested in the narrative and the ways the advertisement draws on the characters and scenario of the game, they enjoy the morphing into game characters and a 'war', they find poetic justice and satisfaction in the boss with his pants around his ankles, and admire the structuring and placement of the advertisement. Those that dislike it singled out the Asian aesthetic and audience, and the lack of 'purpose or reason', and 'the fact that they didn't do a sexy ad'.

The activity provides a space for the articulation of views potentially taboo. The boys' responses show a mix of informed and reflective analysis with more blunt accounts of their observations and what they enjoyed. Despite the apparently feminist agenda presented by the girls' assertiveness and victory, and one boy's observation about sexism, for most of them the women's presence is a more a matter of 'hot chicks [who] drink Coke', and for one at least, it's a matter of regret that 'they didn't do a sexy ad'. They like 'the action, the violence and the fighting'; one boy observing that this is 'good' (in that good reasons for the four to fight have been established in the narrative), but most straightforwardly relish it regardless of the need for such claims. They like the way the characters 'just break out into a war and become these cool monsters of *World of Warcraft*'; that 'they go into battle' and 'they beat the boss up'. The mix of humour, 'action' (violence) and sex (presented as politically acceptable, thanks to girl power) is honed to appeal to this younger end of 'under 18 males'; the game, the humour, and the 'rightness' of the women's cause creating a context in which it is safe in their responses to openly identify qualities that elsewhere in the classroom they might feel obliged to censor or deny (e.g., sex and violence).

The comments made about the Asian orientation of the clip are murkier. It is not clear from the comments here whether they represent a negative response to the 'pan-Asian' aesthetic of the boss's office setting or the game; resistance to being interpellated as an Asian viewer, or indeed, of being the subject of (any) advertiser's targeting; dislike of the game itself; resentment of being forced to study the advertisement at school; in-class feuding; or outright racism. The clip would seem to be a case in point of the ways in which values and ideologies are more readily visible in less familiar texts, or texts where taken for granted values and understandings are not shared by the viewer, thus rendering the marketing aspect of the advertisement more prominent. It is perhaps no coincidence that the student who makes most explicit reference to the effectiveness of the marketing is also the one who takes the most extreme position on its Asian qualities.

In addition to asking what they liked and disliked, we asked students to write an imaginative piece: a report to the marketing manager outlining an advertisement using *World of Warcraft* to sell Coca-Cola. Some students wrote before they had seen the advertisement, and some after. We asked them to include reference to the target audience, to identify specific features of their advertisement chosen to appeal to their audience, the effect they wanted to have on the target audience, and why they chose *World of*

Warcraft. These examples give a sense of the ways the students drew on familiar conventions to come up with new narratives and images:

Warcraft and Coke Ad

A character that is good is getting beaten by evil characters. The only way he can defeat them is if he finds the almighty coke machine gun. It shoots out 5 coke cans a second. He finds it and goes to defeat evil. He defeats evil with coke machine gun and takes a coke can to drink after his efforts.

Advertising

Instead of using potions to heal it could be a coke and that's what makes and gives them all their health back. To defeat the characters you have to shoot a coke can so it opens on the character and then the acid of the can breaks him down and the character then deteriorates. On your journeys you bust open a coke can on the hills of snow. It freezes a path for a quick and easy way to your destination. 'Sometimes coke helps find your path'. Instead of morphing into a creature you can morph into a coke can and rule people.

Selling Coke with help from World of Warcraft

Have same ad (trailer) but:

Before woman changes into animal have her find a puddle of coke and a tipped bottle beside it. Then she gets the ability to change, from drinking coke.

Same idea with skeleton. Have him be a pile of bones that blow away from a gust of wind and end up in a puddle of coke.

What is striking in these three pieces are the ways in which the writers have incorporated and extended images and transformations characteristic of computer games in lively and imaginative ways. The students have no difficulty in seeing such activity as part of marketing, but engage energetically in finding catchy and imaginative images and ideas. They delight in creating 'new and better' ways to sell Coke, demonstrating a kind of commitment or faithfulness to both product and genre reminiscent of the fan culture creativity that games companies sometimes capitalise on seeking 'new and better' versions of their games. Just as Willett (this volume) suggests in her chapter, young people's productions of spoofs and parodies which incorporate existing media structures, as these narratives do, can provide agency in the construction of new meanings and the development of critical stances towards them. Humour, parody and play can work to consolidate identity work and positioning within peer culture, particularly for young men. In this instance, the writing requirement enabled the young writers to fulfil this school-imposed task, and to play with more interesting and contradictory stances in relation to the product at the same time as critiquing it.

The last two pieces show a sophisticated understanding of marketing agendas, and an odd mix of disapproval and celebration in the cleverness of their ideas:

> *The design team and myself have just finished the new ad. This really should encourage people to buy our product. We have incorporated the popular online game* World of Warcraft *into the commercial. This should appeal to our target audience of 12–30 year old men. The main appeal is the game-like action in the ad. It's fast, it's compelling and the teenagers love it. The graphics are nothing short of spectacular. We want the audience to feel as if you can't have one without the other. Warcraft and Coke are one. We want people to feel Coke is the superior soft drink, as in the ad it defeats evil. The reason for using* World of Warcraft *was to catch the audience's attention. They notice Warcraft, they see Coke. Simple. As over 2.5 million play Warcraft in China alone, I think this ad will prove to be very successful.*

> *Our coke sales are running low and we need a new advertisement to boost our recommended sales. Here is an idea that you may take to your liking as the based target audience is young males as many of our sales take place at schools and if we can get our message and our product across the board to these young children then it may help us with the economy. This advertisement is based on the new game of* World of Warcraft *and it is the number one selling game across the globe. It is a real hit among all young males and it is very may be the source we are looking for to get our product across on. Most young males in the world know of this game and if we can incorporate coke into its name we will have a sure selling product. Children are brainwashed with this game, getting addicted to the computer, developing a new personality because of the game and maybe if we can accompany the* World of Warcraft *game with coke it will boost our popularity over other companies. The ad will generally start off with three kids throwing stones at cans of soft drink on a railing trying to knock them off. Suddenly the world starts changing and the kids become characters from the game and instead of stones being thrown it is arrows being shot. The competition gets aggressive trying to knock over the cans and it comes down to the last can, the coke can. They can't hit the can and eventually waste out of arrows. Then a message comes up and says "COKE, KNOCKING OVER THE OPPOSITION BUT AFTER ALL WE'VE BEEN THROUGH WE'RE STILL STANDING TALL AND STRONG THROUGH THE AGES. COKE, AAAAAAAHH-HHHH, LOVE THE TASTE'.*

> *Hope you like it . . . .*

Like the first three examples, the writers here have a strong grasp of the conventions of a range of genres—those of games like *World of Warcraft*, of advertisements, and of this school based task that asks students to fulfil a particular purpose at the same time as writing in role. Both students comfortably adopt the persona of members of the design team, developing a voice and personality consistent with this role but subtly distancing themselves from that character. Their implied criticism of their character's voice and position also works as wry appreciation of the cleverness of the strategy, and the wit involved in developing it.

In the first of these two last pieces, awareness of the expressly manipulative uses to which the game will be put is evident in the reiterated 'we want the audience/people to feel . . . ', and the straightforward explanation that 'the reason for using Warcraft was to catch the audience's attention'. The concept of merging of game and drink in players' minds ('you can't have one without the other), the parallels established such that Coke is regarded as the 'superior' soft drink, and the snappy, one word comment on the effectiveness of the equation—'simple'—suggests the writer's satisfaction with the cleverness and neatness of the proposal, both his own and by extension that of the original WOW/Coke clip that they were shown. This writer is also highly skilled in meeting the demands of this school-based task—that he demonstrate knowledge of how advertising works, and the factors advertisers characteristically call into play. Writing in character, he refers to the target audience (of 12–30-year-old men), explains why the game provides an appropriate focus, and highlights the numbers of players he anticipates his campaign will reach. He constructs and effective and polished argument and meets both the imaginative and the analytic expectations implicit in the task.

The writer of the last piece similarly remains in role while also managing the dual demands of demonstrating knowledge of advertising forms and practices while critiquing them. The character this writer creates is more sinister or cynical than most—games are a good advertising avenue because 'children are brainwashed', they are 'getting addicted' and are 'developing a new personality'. He incorporates appropriate knowledge of market requirements—schools are referenced as useful sites for sales, and arguments advanced for the choice of game. As in most pieces, a scenario is offered where transformation of some kind takes place, with the less common addition of a slogan to take away. In this instance, however, rather than Coke aiding in 'the defeat of evil' (a frequent theme), it is Coke itself that is the winner. There is strong suggestion here, perhaps unintended, that the losers are the very kids who make up the target audience—the kids throwing stones at cans of soft drink 'waste out of arrows', with Coke 'knocking over the opposition' (other drinks or the kids?). Coke remains 'standing tall and strong through the ages'. The capitalised 'message' and concluding comment have the ring of authority and satisfaction of an effective advertisement and a job well done. The writer's critical perspective is achieved through the internal contradictions and the telling coda to the attractiveness proposition—'hope you like it'.

CONCLUSIONS

The advertisement and the boys' responses to it bring into focus points about convergence, glocalisation and digital culture and the complexities of teaching about this with audiences already knowledgeable about parallel processes (computer games, marketing) to some degree. All students rapidly recognised at least some of the mechanisms used within the clip to market the game, but most were ready nonetheless to simultaneously acquiesce or revel in the pleasures the clip entailed. Overlaps of the kind the Convergence Culture Consortium describe are evident in many ways. Many elements of the students' writing directly reflect the creative energy and genre specificity characteristic of 'participatory culture' whereby 'consumers become active participants in shaping the creation, circulation and interpretation of media content', and would seem to have the effect of 'deepen[ing] the consumer's emotional investment in the media property, and expand[ing] their awareness of both content and brand' (Convergence Culture Consortium, 2006, n.p.). One boy joked he was off to get a *Coke* at recess immediately after the clip was shown.

While some, such as the writer of the last narrative, were clearly aware and critical of the ways in which soft drink marketing is at targeted at schools and school aged children (this was at the time when the 'banning' of soft drink sales from schools in the U.S.A. and potentially Australia was being mooted in the press), the general feeling about Coke using *World of Warcraft* as a vehicle to sell the product was one of acceptance, even admiration. For most of these boys, the wit and cleverness of the concept, accompanied by what was effectively a competitive effort to match or exceed this in their own imaginings, far outweighed prosaic condemnation, unless they were able to find ways to do this that did not brand them as 'uncool'. At the same time, however, there was a wry recognition of the phenomenon, and a weighing up of how to position oneself within the context of peer culture to respond. With respect to convergences of 'fan, brand and style culture', in the case of these students, brand and fan culture had less of a role to play, but 'style culture', and their friendships and self-concepts, mattered considerably.

In seeking to understand the boys' experience of, and responses to, the advertisement, it is not enough to assume that enjoyment equates with uncritical acceptance, nor that the students were naively acquiescent or unaware. In most instances, their responses show a complex mix of recognition of marketing strategies, the use of in-game narratives, aesthetics and allegiances, and the compelling power of the invitation to participate and be 'cool'.

Even though they were not Warcraft players, the clip caught and kept their attention through its humour, graphics and story line, but also through its 'action', the girls and the desire to be 'cool'. Girl power ideologies meant it was safe to speak about their enjoyment of the presence of 'hot chicks', in ways that seemed not to trouble them or present them with contradictions or complexity. When pressed in interviews, most argued it was good to barrack for the girls, and that they were a big part of the advertisement's appeal. The narrative, with its invitation to side with the women in a satisfactorily justified battle with the ogre, coupled with humiliating the boss with his pants around his ankles, seems closely targeted to appeal to boys of this age. Only one boy in the writing collected identified 'a bit of sexism', as part of a larger set of things he didn't like about the game.

The global media landscape provides rich opportunities for potential convergences between advertising, youth culture and computer games, and the co-option of ideologies within and outside the game to create new marketing opportunities. The *World of Warcraft* Coke advertisement was designed for television, and despite its presence on the Warcraft community website, was not intended as part of game play. It's a small step, however, to imagine how it, and others like it, might become so. Given the pervasiveness and attraction of such online worlds, and the sites they offer for all manner of 'real world' engagement and meaning making, there are strong arguments both for understanding more about the kinds of textually mediated experiences, negotiations and relationships young people make and encounter there, and for equipping young people with informed and critical perspectives on forms of marketing and interpellation they might experience in online textual worlds like these. Teenage boys are already part of the global media context, already familiar with many of the images, opportunities and pleasures that affiliation with this provides. Classroom work that seeks to address this complexity needs to provide spaces for both analysis of the dimensions of effective advertising and recognition of the attractiveness of digital culture and the global cultural industry. Teachers and researchers concerned with digital culture and the nature of the worlds and relationships young people encounter there need to learn more, themselves, about the ever changing dimensions of landscapes such as these.

ACKNOWLEDGEMENTS

Thanks to the teachers, James and Tim, the Year 8 students and the school, for their generous participation in this project. Thanks, too, to Claire Charles for her assistance throughout, and insightful feedback on earlier versions of this paper.

NOTES

1. Pseudonyms for the teachers are used throughout.

REFERENCES

Alvermann, D. E., Moon, J. S., & Hagood, M. C. (1999). *Popular culture in the classroom: Thinking and researching critical media literacy.* Newark, DE: International Reading Association and National Reading Conference.
Buckingham, D. (Ed.). (1998). *Teaching popular culture: Beyond radical pedagogy.* London: UCL Press.
Buckingham, D. (2000). *After the death of childhood: Growing up in the age of electronic media.* Malden, MA: Polity Press.
Buckingham, D. (2003). Media education and the end of the critical consumer. *Harvard Educational Review, 73*(3), 309–328.
Burns, E. (2006). Marketing opportunities emerge in online gaming venues. *Clickz.com,* August 1, http://www.clickz.com/showPage.html?page=3623035 Retrieved August 1, 2007.
Consalvo, M. (2006). Console video games and global corporations: Creating a hybrid culture. *New Media and Society, 8*(1), 117–137.
Convergence Culture Consortium, Massachusetts Institute of Technology. (2006). About C3. Retrieved July 19, 2006, from http://www.convergenceculture.org/aboutc3/index.html
de la Ville, I., & Durup, L. (2008). Achieving a global reach on children's cultural markets: Managing the stakes of inter-textuality in digital cultures. In J. Marsh, M. Robinson, & R. Willett (Eds.), *Play, creativity and digital cultures.* London: Routledge 36–53.
De Mooij, M. (2001). Convergence and divergence in consumer behaviour: Implications for global advertising. *International Journal of Advertising, 22,* 183–202. Retrieved July 19, 2006 from http://www.mariekedemooij.com/articles/demooij_2001_admap.pdf
Facer, K., & Williamson, B. (2004). More than 'just a game': The implications for schools of children's computer game communities. *Education, Communication and Information, 4*(2/3), 253–268.
Gee, J. P. (2003). *What videogames have to teach us about learning and literacy.* New York: Palgrave Macmillan.
Hopkins, S. (2002). *Girl heroes: The new force in popular culture.* Annandale, NSW: Pluto Press.
Jenkins, H. (2004). The cultural logic of media convergence., *International Journal of Cultural Studies, 7*(1), 33–43.
Kress, G. (2000). A curriculum for the future. *Cambridge Journal of Education, 30*(1), 133–145.
Lindstrom, M. (2007, June 11). Are you ready to play? *AME Info.* Retrieved December 12, 2007, from http://www.ameinfo.com/123133.html
Luke, A., Freebody, P., & Land, R. (2000). *Literate futures.* Education Department, Brisbane, Australia.
Madden, N. (2005). *Coke* brings fantasy to life: Summer program appeals to teens with music and games. *AdAgeChina.* Retrieved July 19, 2006, from http://adage.com/china/article.php?article_id=46112
Marsh, J., & Millard, E. (2000). *Literacy and popular culture: Using children's culture in the classroom.* London: Paul Chapman.
Ministerial Council for Education, Employment, Training and Youth Affairs. (2005). *Statements of learning for English (NSLE) 2005.* Carlton South Victoria,

Curriculum Corporation. Retrieved October 5, 2007, from http://www.mceetya. edu.au/verve/_resources/SOL_english_06.pdf

Morely, D., & Robins, K. (1995). *Spaces of identity: Global media, electronic landscapes and cultural boundaries*. London and New York: Routledge.

Nixon, H. (2003). New research literacies for contemporary research into literacy and new media? *Reading Research Quarterly, 38*(3) 407–413.

Robertson, R. (1995). Glocalisation: Time-space and homogeneity-heterogeneity. In M. Featherstone, S. Lash, & R. Robertson (Eds.), *Global modernities*. London: Sage. pp. 25–44.

Schaffer, D. W., Squire, K. R., Halverson, R., & Gee, J. P. (2005). Video games and the future of learning. *Phi Delta Kappan, 87*(2), 105–111.

Sefton-Green, J. (Ed.). (1998). *Digital diversions: Youth culture in the age of multimedia*. London: UCL Press.

Steinkuehler, C. A. (2004). Learning in massively multiplayer online games. In Y. B. Kafai, W. A. Sandoval, N. Enyedy, A. S. Nixon, & F. Herrera (Eds.), *Proceedings of the sixth international conference of the learning sciences* (pp. 521–528). Mahwah, NJ: Lawrence Erlbaum Associates. Retrieved July 19, 2006, from http://website.education.wisc.edu/steinkuehler/papers/SteinkuehlerICLS2004.pdf

Turkle, S. (1995). *Life on the screen: Identity in the age of the internet*. New York: Simon & Schuster.

Universal McCann. (2006, June). *Trend Marker: Parallel worlds: The impact of massively multiplayer online games* (MMOGS). Retrieved December 12, 2007, from http://www.universalmccann.com/page_attachments/0000/0016/Parallel_Worlds.pdf

Yee, N. (2005). The psychology of MMORPGs: Emotional investment, motivations, relationships, and problematic usage. In R. Schroeder & A. Axelsson (Eds.), *Avatars at work and play: Collaboration and interaction in shared virtual environments*. London: Springer–Verlag. pp. 187–208. Retrieved July 18, 2006, from http://www.nickyee.com/daedalus/archives/02_04/Yee_Book_Chapter.pdf

Yee, N. (2006). The demographics, motivations and derived experiences of users of massively multiplayer multi-user online graphical environment. *Presence: Teleoperators and Virtual Environments, 15*, 309–329. Retrieved July 18, 2006, from http://www.nickyee.com/daedalus/archives/pdf/Yee_MMORPG_Presence_Paper_pdf

3 Achieving a Global Reach on Children's Cultural Markets

Managing the Stakes of Inter-Textuality in Digital Cultures

Valérie-Inés de la Ville and Laurent Durup

THE GLOBAL REACH OF RECENT FRENCH ANIMATION PRODUCTIONS

Some Evidence[1]

Today, the French animation industry is ranked third in the world and first in Europe, producing around 300 new hours of animation each year. French producers benefit from efficient governmental subsidies designed to support creativity in this industry. Partly due to this unique and enviable system, since the year 2000, French productions globally have received more than 50% of their budget from national sources. For example, in 2005, on average, 17% of the budget for an animation series came from French producers, 26% from the French TV channel which bought the series, 16% from governmental subsidies, 5% from 'pre-selling' to other French TV channels,[2] for a total of 64% of the budget.

This financial scheme for French animation production reflects a significant change since the year 2000. At the end of the 1990s, 'foreign initiatives' funded 30%–40% of French animation production, whereas in 2004 it was only 5%, and 13% in 2005.[3] In 2000, coproductions with foreign countries represented 35% of the average budget, and pre-sells to foreign countries 9%; in 2005, foreign coproductions were only worth 19%, whereas foreign pre-sells were worth 14%, increasing the part of French ownership over the animation programmes. And between those two periods, a crisis happened that might have changed the deal.

From 2001 to 2004, the animation industry was undergoing a deep crisis, coinciding with a glut in the audiovisual distribution market, which was partly due to a more generalized economic crisis as well as to an excessive number of products on offer, leading to a situation of price reductions and aggressive competition between producers on international markets. In that context and more than ever, the U.S. market was considered as a kind of 'Eldorado': the third in size, after China and India, but with a more globally recognized culture and uses for business closer to the European ones. This situation entailed two major consequences for French producers: concentration of companies and internationalization of their market.

From 2000 to 2005, a hundred societies were producing animation in France. Nevertheless, only four of them produced at least one programme each year, and the 15 biggest of them have made 69% of the animation hours produced by the profession. Meanwhile, Media Participations bought Storimages, Ellipsanime and Dupuis Audiovisuel, creating a group that produced 16% of French animation production from 2000 to 2005. Antefilms bought France Animation to form Moonscoop SAS (8% of French animation production from 2000 to 2005). And the new group Finhera was created, merging Marathon Medias and Tele Images Kids, for a total of 10% of French animation production from 2000 to 2005. Thus, in only 5 years, three groups emerged and produced a third of French animation, leading to a concentration situation never known before.

Meanwhile, owing to the success of some series like *Totally Spies* or *Funky Cops*, sold in more than 100 countries, French producers proved their capacity to capture the global market. Although representing only 10% of audiovisual French production in 2004, the animation genre was worth 55% of presells and 37% of sells of audiovisual French production to foreign countries, and 50% of coproductions.

In 2004, of the €22.8 million pre-sales to foreign countries, €1 million included pre-sales to the US; and of the €39.5 million sales in 2004, €3.4 million were sales to the US, although it was only €2.2 million one year before.

Animation is worth 65% of French audiovisual exports to North America overall: Therefore, the American 'Eldorado' no longer appears unreachable. How did French productions seduce an American audience? Part of the answer, paradoxically, lies in the French process of fund raising.

Fund Raising: A Dual Construction

As described above, the French TV channel that bought the first exclusive broadcasting rights (referred to as 'the main broadcaster' in the following sections) is an important actor in the animated series fund raising process, contributing an average of 26% of the budget, that is, more than the producer itself. Given this situation, an animated series has virtually no chance for development until the main broadcaster is found, a position that greatly influences the cultural content in favour of the main broacaster. Very often, main broadcasters ask for important changes in the series before signing. But even more often, the series are designed for a specific channel, and elements are included to attract a particular channel and reinforce its competitive positioning and its identity within the saturated media marketplace. But to achieve commercial success in the United States, the producer also needs to take into consideration the requests of American broadcasters. For example, when Tele Images Kids wanted to sell a TV series to Cartoon Network, it created *Atomic Betty* with a graphic style very close to Cartoon

Networks' one. For the audience, it was virtually impossible to differentiate the French series from the other series on the channel. So how is it possible to seduce two broadcasters airing in two different cultural contexts? In fact, it is far easier than it seems.

Channels know their own audience perfectly: They know the gender make-up, age-range, programme preferences, and so on. To please their audience, channels tend to offer series that match 'revealed tastes'—series that draw on elements already used in previously successful series. By doing so, channels take fewer risks, but even if they lower innovation, they are sure to respond to present children's culture. Thus, if we take a look at broadcast animation on French channels in 2005, we see that only 38% of it is French. Reflecting children's globalised media culture preferences, smaller but nevertheless important parts of TV series aired in France are from the United States and Japan.

As far as U.S. channels are concerned, 1998 was an important year leading to major changes: In 1998, *Pokémon* appeared in a favourable context and became one of the most successful series in U.S. animation history. Thus, soon after, Japanese animation represented about 33% of the animation being broadcast on Cartoon Network. Drawing on the success of *Pokémon*, 'animé' became a standard fully recognized by audiences in the United States. As a consequence, meeting Japanimation standards is considered an additional competitive advantage for reaching U.S. audiences.

Considering these observations, we can say that children in the United States and France share a huge common cultural ground based on Japanese and American cartoons. Child audiences have 'revealed' some common tastes, and therefore it is possible to build a budget with a French main broadcaster, while also including U.S. coproducers or prebuyers. Once the series is under production, the U.S. coproducer acts as a gatekeeper, opening commercial contacts to sell the series in the United States.

To understand the nature of this emerging realm covering a global animation culture which is common to child audiences, as well as to explore the stakes involved in French producers' use of this culture in their efforts to reach the international global marketplace, we develop our analysis through the concept of inter-textuality.

Inter-textuality: Playing with Pre-existing Standards

This section consists of a brief reminder of the main definition given to inter-textuality in literary analysis. Dealing with this notion leads to an understanding of writing as a permanently creative flux that integrates previous standards and conventions in order to produce texts likely to be readable, understandable and recognizable by an audience.

In a nutshell, inter-textuality has been introduced by Julia Kristeva (1980) as a reaction to the tendency to analyze texts as discrete and close units whose meaningfulness lay in their internal structure. Drawing on the dialogical perspective introduced by Mikhail Bakhtin (1968) in literary theory, Kristeva contends that texts become meaningful if considered as a fragment in connection to former texts. Shared codes allow both the writer and the reader to recognize, situate and appreciate the text in the continuum of literary production.

It is worth noting that this post-structuralist perspective sees every text as dominated by previous texts that impose a universe of codes in relation to which it will be read and understood by several audiences. This entails a drastic shift in the analysis of writing and reading by focusing on the process of structuration through which the text came into being. By questioning the romantic roots that lead to the invention of the notion of 'authorship', this perspective lays special emphasis on the fact that, to communicate with an audience, writers are compelled to use pre-existing concepts and conventions. When speaking they are at least partially spoken, and their individual creative skills are socially founded in shared language and scriptural conventions. This is why Roland Barthes defines the text as a tissue of quotations, a creative art involving the weaving of former codes, references and genres (Barthes, 1974).

Texts draw upon a large range of codes and social norms that allow them to be assigned to a particular genre. Genres are situated and evolving conventions that make it possible to classify texts and frame their relation to each other. Literary theorists have brilliantly demonstrated that the definition of genres is quite fluid and is subject to ongoing changes and social renegotiations, leading to a permanent blurring of borders and a constant mitigation of their distinctive characteristics. Thus, categorizations are always precarious and need to be revised constantly to stick to literary creative evolutions: new genres such as advertorials, advergamings, infomercials, edutainment, docudrama, etc., are always emerging in the global digital cultural landscape.

Although the notion of inter-textuality has been largely criticized as being too broad, it is worth understanding inter-textuality as a principle that needs to be adapted to different situations. Thus, Gerard Genette (1992) suggests distinguishing between five kinds of 'textuality':

- *Inter-textuality*: operates within a text through quotations, plagiarism, allusions;
- *Para-textuality*: includes all the peripheral elements that closely surround the text and help situate it such as titles, headings, dedications, footnotes, prefaces, illustrations, etc.;
- *Archi-textuality*: the twofold process, activated by the author and the readers, by which a text is designated to form part of a literary genre;
- *Meta-textuality*: implicit or explicit criticism of former texts included in the text itself;
- *Hyper-textuality*: the transformations of former codes or genres to which it relates that the text carries out.

Moreover, in the context of digital cultures, some authors (e.g., Chandler, 1995) have emphasized profound disruptions in texts' conventional linearity that hypermedia navigation allows. Some exploratory displacements appear possible through structures such as simultaneity, argumentative shortcuts, coexistence of textual fixity and visual mobility or vice versa, leading to meaning being conveyed in very unusual forms and to the enhancement of truly innovative reading processes.

We are fully aware that such a brief presentation of the complexities linked to the notion of inter-textuality and its use in cultural studies and literary analysis is a risky undertaking. Nevertheless, in the context of this chapter, we aim to use this notion as a tool in order to understand the processes of production involved in children's global media texts.

DIFFERENTIATING LAYERS OF INTER-TEXTUALITY IN SUCCESSFUL FRENCH ANIMATION SERIES

Our attempt to use the notion of inter-textuality to animated TV series production draws on the analysis of two French TV series that have both achieved global success in children's cultural markets: *Code Lyoko* and *Totally Spies*. When contrasting these two French series, another interesting dimension appears: the fact that they privilege two contrasting cultural contexts, a Japanese one for *Code Lyoko* and an American one for *Totally Spies*.

Methodology

The methodology we used to explore the various dimensions of inter-textuality that could be traced within these two French series involved different techniques of direct and indirect analysis.

Firstly, by viewing several episodes of the series unfolding over different seasons, we undertook a two dimensional direct analysis which covered:

- A narrative analysis of plot construction: This included an analysis of the storyline, the way the main characters of the series behave, the quest they collectively follow, and the topics of interest they deal with, the cultural elements they refer to in their talks and the decisions they make. Plot construction refers to previous genres and stories known by the audience, and it easily reveals evidence of 'inter-textual' references the audience is supposed to master.
- An assessment of graphic and artistic choices: This included a close examination of the evolution of the characters themselves—their progressive ageing, the way they are dressed, the accessories and devices they use, etc. It also included the different settings and the rendering techniques privileged according to the nature of the scenery (e.g., realistic or fantasy).

Secondly, we did an analysis of marketing and managerial statements made by the companies through their official declarations within professional circles (press, magazines, websites, commercial documentation, etc.) and in direct interviews with the authors and producers of the series.

Thirdly, we made a 'loose survey' of the reactions of the audience by obtaining from marketing managers a selection of the most significant direct contacts they had by mail, and by exploring, through several wordings related to inter-textuality, the discussions going on amongst fans within the official online forums of the series. The official websites are massively supported by active fans, and they develop over the different seasons covered by the series, allowing researchers achieve a deep understanding of the evolution of the topics of interest for the community of fans.

Totally Spies

The series *Totally Spies* was designed to reach primarily an American audience as a coproduction between the French company, Marathon, and the U.S.-based Fox and ABC Family (Disney group) associated with Cartoon Network (Warner Bros). Although created by French authors, this series has a totally U.S. setting (a campus in Beverley Hills), and its scripts were designed in accordance with U.S. taste and references in order to convince the audience that it was an U.S. production. It seems to us that a wide part of the originality of the humour of the series is based on a stereotypical European vision of American references. By this means, the series achieves a double positioning: fully reaching, for the rest of the world, the 'U.S. Standard', whereas appearing as original and innovative to the U.S. audience. 'We produce French series disguised in American ones', comments Vincent Chalvon-Demersay, cocreator of *Totally Spies* and General Manager of Marathon Animation.[4]

Totally Spies plays on both U.S. and Japanese references. The most evident references on which *Totally Spies* plays are, for instance:

- *Charlie's Angels*
- *Beverley Hills*
- *The Avengers*
- *James Bond* (gadgets)
- Japanimation's standards in graphical attitudes (exaggeration of the expressions, smiles)
- *Cat's Eyes*[5]: close fitting clothes including high heeled shoes, trio of women with a secret life

Stéphane Berry, cocreator of the series and artistic director of Marathon Animation says:

> The style of *Totally Spies* is a melting between the American style, which associates action and comedy, and Japanese design for the aesthetic

Figure 3.1   TS © Marathon, Mystery Animation Inc.

environment and the emotions expressed through the large eyes of the characters. It is an entertainment product that didn't exist before in the French audiovisual landscape and that defines a new genre of animation whose uniqueness and originality is appreciated by the audience.[6]

Episodes from the series have been met with great enthusiasm in 130 or so countries, in the 6- to 11-year-old audience age bracket. However, the production was risky in at least two ways: targeting girls with a genre generally aimed at boys (adventure and comedy), and using animation, given that girls of this age usually have shifted to live shows. The risk was worth taking, and the result was obvious: girls were attracted to the series but boys also appreciated it. On a smaller scale, *Totally Spies* did what *Pokémon* had done years before: reaching a wide audience in terms of age range and gender. After more than 130 episodes aired, the fifth season is under production, and the three heroines, who are getting older and are now fully mature, are placed in a new setting—the university.

Totally Spies is graphically conceived as a brand that can be used on a wide variety of merchandise, and the objects used in the series have their own design bible[7] (e.g., colours, use of the *Totally Spies* flower, spy gadgets looks like everyday objects—powder, hair dryer, lipstick, back pack, etc.).

Code Lyoko

More recently, another French series, *Code Lyoko*, has also been broadcast on Cartoon Network in the United States, where, in the space of a year, the MIGUZI block[8] in total scored excellent audience ratings. In the *Code Lyoko* series, planet Earth and the parallel universe, called Lyoko, face the threat of annihilation. A super virus has infected the central processing units, called X.A.N.A., and only four kids can foil the mad computer's evil designs. The heroes of the series, Yumi, Ulrich, Odd and Jeremy, lead double lives: ordinary students in their early teens in everyday life, action heroes in the virtual digital world when X.A.N.A. attacks. Intended for 6- to 14-year-olds, the series has proven most popular amongst 8- to 10-year-olds on Cartoon Network.

The narrative itself combines two plots corresponding to the two settings: a classic school show, based on groups of friends, love rivalry, pupils' everyday lives similar to the audience's, and an action adventure intrigue in a virtual world where magical objects make everything possible. In *Code Lyoko*, 'Science-fiction blends with sitcom in a cross-genre'.[9] This narrative construction has already been proven effective in the *Power Rangers* series, whose more than 20 years longevity is an exceptional phenomenon.

Another characteristic of the plot construction is that it has been designed to reach two targets: an action / adventure genre that targets mainly boys, blended with a sitcom genre dealing with love relationships, friendship, rivalry, and exploring a wide rage of feelings amongst teenagers, which overtly targets girls. Even though girls usually watch action series aimed at boys, the risk in this case consisted of convincing boys to take an interest in the sitcom part of the series.

In the real world of *Code Lyoko*, everything is quite mundane: Jeremy's exceptional skills in programming are viable, and the background is a typical French junior high school. The sitcom part of each episode uses 2D animation techniques that constitute a traditional standard in animation aimed at kids. However, high technical quality animation with 3D renderings is used for half of each episode (corresponding to the virtual world). Full 3D is sometimes avoided because of its unreal texture, but here the script is strongly linked to video games, therefore this choice becomes totally significant and consistent.

As far as graphic design is concerned, in the virtual world of Lyoko, Japanese references prevail, both in the style of the drawings and in the objects and the costumes used by the heroes (who transform into warriors). Yumi, for instance, wears a kind of short kimono and uses a fan, a traditional

Figure 3.2   Code Lyoko™ © Moonscoop—France 3-2007—All Rights Reserved.

weapon of samurais. Other weapons, such as katanas (Japanese swords), or magical techniques, such as duplication, are archetypal of Samuraï or Ninja fighting arts. Lyoko is also a direct reference to *Digimon's* and *Matrix's* worlds. In those universes, there are always two worlds, the 'real' and the 'virtual' one, and there is a need to protect the 'virtual' world to save the 'real one. The reference creates a new layer of inter-textuality that moves closer to dominant plots and graphics used in many action video games.

Moreover, fashion (seen as a strong interest for tweens and adolescents) has been carefully taken into account. The heroes are very well dressed and use up-to-date accessories (latest generation of mobile phones, designer handbags, earrings, belts, piercings, etc.). This focus on fashion also opens avenues to launch spin-offs in that area.

When developing animated TV series aimed at 6- to 12-year-old children, production managers implement a dynamic of inter-textuality that allows an integration of the TV series into the global digital cultures children appreciate, recognize and master. Children's cultural background is partly shaped by the media saturated global marketplace and includes TV animated series, as well as video games and websites. Even if television stands out as the preferred media of young kids, they also know how to use computers, read magazines, listen to the radio, record music, send e-mails, send SMS (text messages) on mobile phones, surf on the Internet, and so on (Kline, Dyer-Witheford, & De Peuter, 2003; de la Ville, 2007). Thus the stakes of inter-textuality spread far beyond a specific media to offer an encompassing perspective of all the media that children learn to use in the course of their development.

MASTERING THE STAKES OF INTER-TEXTUALITY

The notion of inter-textuality allowed us to better understand the strategies consciously or unconsciously set up to bring *Totally Spies*, *Code Lyoko* and other French animated series to an international audience. We will now see that although the five kinds of textuality can apply to those strategies, some of them are more relevant than others when examining the process of production.

Privileging Kinds of Textuality According to Design Objectives

Firstly, inter-textuality gathers various kinds of references, from the most unconscious, not even seen by the creator, to the most conscious, designed not only to be seen but purposely identified as a reference by the audience. As we said, allusions to *Cat's Eyes* may be identified in *Totally Spies*, but can not be considered as a key reference for the audience, due to the fact that *Cat's Eyes* was broadcast in the 1980s and is no longer known, even only its name, by today's kids and teens. This reference is no longer relevant, and we may even wonder if it was intended by the authors of *Totally Spies*, who are also influenced by their own child culture. 'Fan-service' references, on the other hand, are more clearly intentional inter-textual references aimed at particular audiences. 'Fan-service' is an expression coming from animé fans and refers to situations in which the series intentionally adds elements that are not necessary for plot construction, but are intended to please the fans. Such elements are mainly divided into two categories: showing the female heroes in underwear and, what is more interesting for this section, making direct allusions to other TV series. These allusions are purposely designed to be seen by a part of the audience as a way for the producer to acknowledge its fan subculture, and to give that part of the audience an opportunity to impress the other part of the audience. This tradition of 'fan-service' was usually aimed at teens and adults, but is gradually reaching a younger audience. *Shrek I* and *Shrek II* are feature films filled with 'fan-service' references, as for example in *Shrek I* the upside down kiss scene, which references *Spider-Man*, or in *Shrek II*, the axe in front of the moon, which references the *Batman* sign.

Secondly, 'para-textuality' although relevant, is a marginal tool, given that peripheral elements (e.g., titles, headings, dedications, footnotes, prefaces) are far less numerous for animation than for written texts. We could use the concept of para-textuality to say that an animated series must have a title that relates to its content, which seems quite obvious. However, para-textuality indicates more than just a reference to content, but also meanings associated with particular content. For example, when the pilot episode of *Code Lyoko* was presented to potential partners, the series' name was not *Code Lyoko* but *Garage Kids*. Both names are direct references to the content of the series, but one refers to 'the code to save the real world in a videogame-like universe' and the other to the mundane building where the door to the 'virtual' world is located. And actually *Code Lyoko* was more appealing for a global audience than *Garage Kids*.

Thirdly, 'archi-textuality' is very important for the audience. Archi-textuality opens a set of possibilities in order to blur genres and references, thus creating some ambiguity actually felt by a part of the audience. For instance, a young boy from the *Code Lyoko* community sent the following email:

> There is this forum on the site, and people are arguing if *Code Lyoko* is an animé or not. Animé is Japanese cartoons, but some people think that *Code Lyoko* is a French animé. What is it? An animé or not?'[10]

If the existence of a French animé is impossible for this young boy, the debate about whether or not *Code Lyoko* and other French productions could be considered animé is a sign that animation genres are strong concepts for child audiences, strong enough to categorize even the 'aliens', as *Code Lyoko* might be considered, but also strong enough to lead some children to fight to preserve their purity. But archi-textuality is not only the result of the readers action, and there is still room for the authors or producers to make this classification evolve gradually, in order to alter its categories or create new ones in which French productions integrate more easily in international children's animation culture, and therefore on the digital global marketplace.

Fourthly, 'meta-textuality', considered as implicit or explicit criticism of former texts included in the text itself, holds an ambiguous position, whether we consider that the 'text' is the animation content in a strict way (story, plot construction, characters, and so on) or not. Indeed, within an animated series, criticism of other animation content is usually only done through 'parody' or 'fan service'—considered as an inter-textual reference. Moreover, if reaching the global market implies the necessity to focus on 'common tastes', it could be quite dangerous to negatively criticize the audiences' common tastes. Therefore, the only criticism is a positive one, which weakens the strength of such criticisms. Nevertheless, taking a wider definition of 'content' changes the analytical perspective. Indeed, technique is a strong meta-textual characteristic, as an animated series is also defined by its graphical prowess. Thus, techniques from older series which are obsolete are implicitly criticized when more up-to-date graphical techniques are used by newer series. This is especially true for 3D animation, a graphical style which is highly dependent on computer know-how.

Fifthly, 'hyper-textuality' is probably one of the most powerful tools available for producers, as it allows the use of highest level of references, apart from sequels or plagiarized texts. A fan of *Code Lyoko* expresses quite well the transformation of former codes or genres he identifies as achieved by the series:

> *Code Lyoko* is the greatest show I have ever seen from France. It is neither an *animé* nor a cartoon, yet I love it all the same. I am 17 and from the USA. The characters are unique and special with each displaying a new angle in personality. Even the minor characters in the show (Sissy, Jim, etc.) add to the total genius buffet within the show. I have, in my spare time, written an episode idea for *Code Lyoko* and will be e-mailing it to production director, Joanna Ruer. Please tell me what you think of it. Thank you for the hard work and the hard work of all the people to make this show.[11]

In addition to the appeal of using codes and genres from other texts, hyper-textuality allows targeting of a wider audience, as most of the examples using this technique are not based on other series, but more fundamentally on a global common cultural background. For example, *Ulysses 31* is a series adapting the Odyssey story to a futuristic world in which Ulysses is searching for a way back to Earth, rather than a way back to Ithaca, after killing the Cyclops on his own planet and angering the Gods. It is also worth noting that hyper-textuality avoids being in the shadow of another series in which inter-textual, archi-textual and meta-textual references find their origin. Indeed, for a producer and a series, quoting another programme implicitly or explicitly is a way of reinforcing the notoriety of its own competitors. The proximity of *Digimon* to *Pokémon* is surely what made *Digimon* strong from the very beginning, but also short-lasting.

Even though these different kinds of textuality help clarify specific stakes producers have to face in order to reach a global audience, they do not give a perspective on one of the crucial elements for production managers: trans-media references. We will attempt in the following section to develop more precise tools to explore the specific opportunities of inter-textual strategies within animation productions.

Towards an Integrated Inter-textual Strategy to Achieve a Global Reach in Children's Cultural Markets

As far as series' revenues are concerned, consumer products departments play a growing role in designing long-term policies to promote the most successful series as brands. For instance, in the United States, *Code Lyoko* is turning into a cult phenomenon—its dedicated website receives over 5,000 messages daily featuring children's drawings or suggestions for sto-ryboards. 'There is no doubt that spin-off sales are going to rocket as soon as the products are available', says Axel Dauchez, General Manager of Moonscoop.[12] This grants larger revenue streams for French producers able to reach U.S. or global networks successfully.

Notwithstanding, after four or five seasons, there is an actual risk of boring the audience if some additional events are not successfully planned. Thus, French producers have now to meet new challenges:

- Organizing the progressive escalation in symbolic added value every time a spin-off is negotiated and launched;
- Increasing entertainment added value each time a connected cultural activity is offered;
- Structuring their marketing department to meet the requirements of powerful retail partners;
- Conceiving of complex promotional events involving a large number of partners and licensees;
- Mastering planning and deadlines in the scheduling of different pro-motional events.

This reflection on inter-textuality affords a new perspective as it allows us to differentiate it from the localization process. In managerial terms, localization consists of a downstream process whose objective is mainly to adapt promotional techniques to diverse consumption habits in several countries. Inter-textuality sheds light on the formation of children's cul-tures in a global marketplace as a complex intertwining of cultural refer-ences partially shaped by commercial strategies (competitive positioning of channels' identity, programming choices, spread of spin-off policies, bounded promotional campaigns, etc.).

Drawing on the work by Gerard Genette (1997), who has proposed 'trans-textuality' as a more generic term than inter-textuality, it is interesting to define with more accuracy the different forms of inter-textuality and their associated advantages and risks. We suggest a very tentative classification of different stakes linked with inter-textuality in children's global cultural industries as follows:

- *Media inter-textuality*: refers to the allusions to former genres and codes included in the product itself to help position it in a given medium context (TV series, video game, feature film, publishing, etc.) while making its singularity also perceptible. It includes the dominant genres and codes that are used to situate a cultural production within one specific medium. The audience will appreciate a TV show, for instance, by directly comparing it to previously successful TV series that shaped criteria or codes concerning what is a 'good' or 'enjoyable' show in a specific cultural context. This media inter-textuality makes it possible to appraise the competitive landscape in which a new product will develop itself and to better argue the originality of its positioning with the aim of bringing innovative products to targeted broadcasters. The risks associated with this kind of inter-textuality lay mainly in the efforts carried out to build awareness in children's topical cultural references.
- *Trans-media inter-textuality*: covers the 'translations' to be mastered when a cultural world, aimed at a primary target on one medium, expands through different media which have their own media inter-textual dynamics. Given that media convergence represents a reconfiguration of media power and a reshaping of media aesthetics and economics (Jenkins, 2006 p. 34), trans-media inter-textuality becomes highly complex when it is developed across a large range of media, running from wireless-connected mobile game devices, home consoles, TV series, internet websites, collectible cards, books, comics, toys and even fashion accessories or fabric collection. The difficulty here is to situate the inter-textual dynamics of a medium in relation to those driving creativity in other media. Each industry has its own history, and customers or audiences have developed their own references and codes that need to be cautiously woven together in order to establish the grounds for consistent use of the product. Kline et al. note that 'in the oligopolistic market of the third millennium, branding platforms and establishing synergistic connections with other branches of youth culture becomes essential for cultivating, consolidating and expanding a loyal user base' (Kline et al., 2003, p. 220). In this perspective, the risk lays in the ability to adapt a product from one medium to another, a complex process which sometimes leads to a global reconception of the main characteristics of the heroes or the plot construction to better fit the tastes and comparative elements mastered by the newly targeted audiences.

- *Branding inter-textuality:* expands the symbolic reach of a product used as a brand by developing bounded spin-offs through different media. The managerial stake here consists of building the symbolic synergies necessary to strengthen the appeal of the branding strategy linked to a product. Companies attempt to capitalize on the recognition and symbolic value of their characters by 'repurposing' them into licensed products, which opens new revenue streams. Such 'brand extensions are no longer adjuncts to the core product or main attraction; rather, these extensions form the foundation upon which entire corporate structures are being built' (Klein, 2000, p. 148). Nevertheless, driving this semiotic escalation is a risky task as it increases the saturation of children's cultures with branded products. The complex interplay of codes through time may end in incoherencies, ambiguities, contradictions or omissions. Indeed, pushing bounded promotional policies too far to impose a product in several entertainment activities or media environments can be counterproductive, leading sometimes to indifference and even resistance from the consumers and the targeted audiences. As Kline et al. describe:

> The management of demand for symbolic goods in the post modern market place has been far from flawless, consistent, or uniformly effective, especially as media channels have expanded and cultural markets have become saturated with commercial signs. Rather, the marketers' efforts in some ways increase the very uncertainty they seek to control by creating ever-more sophisticated and jaded young customers. Fads pass, styles change, values realign, markets mature, and boredom and overload increase with saturation and repetition, making management of symbolic value a very risky venture. (2003, p. 237)

Inter-textuality expresses itself in multiple forms and degrees of combinations with different references and codes which are mastered by children. Even though it blurs boundaries amongst animation genres and blends cultural references, this art of hybridization (Consalvo, 2006) is what enables producers to reach a global child audience. Exploring in detail how this hybridization is practised by French, Japanese or American producers might reveal subtle differences in the creative and managerial competencies that are mobilized to reach a global audience. These synergistic connections we have described, that are based on inter-textual movements, are becoming vital for success or even survival in the global animation market, in the game market and even in the toy market, as well as in the internet services market.

Nevertheless, according to Joseph Tobin, the global success of *Pokémon* 'came at a price: *Pokémon's* producers decided that if *Pokémon* were to make it globally, it would have to reduce what Koichi Iwabuchi calls its "cultural odor"‛ (Tobin, 2004, p. 261). Inter-textuality can also be used as a means to blur idiosyncratic references and to refer to 'deodorized genres' (science fiction, for instance) in order to ease the exportability of the production. But in the case of *Code Lyoko*, the episodes have not been adapted for a North American audience and the choice of maintaining a highly situated context (an old Renault factory in Paris, a high school in Sceaux, a suburb of Paris) can be surprising, as it would have seemed preferable to refer to a plainer environment to reach foreign audiences. Nevertheless, there is evidence that it works: North American fans of *Code Lyoko* have investigated the French school system, and they share this information to develop a deeper understanding of what is at stake in the series.[13] Voluntarily or not, this process might, in the long run, add French references to international standards, opening up the global market to productions clearly identified as French.

CONCLUSION

As proposed in literary theory, textual representations are never pure and simple, and they are very closely connected to systems of power in society. As we can see in relation to children's global media culture, the commodification of culture leads to the production and marketing of experiences. 'Play is what people do when they create culture' and 'the commodification of cultural experience is above all else, an effort to colonize play in all of its various dimensions and transform it into a purely saleable form' (Rifkin, 2000, p. 260). Young consumers become active participants in shaping the creation, circulation and interpretation of media content. Thus, the techniques of experiential marketing pervade brand extension and marketing communication, which are managed through an encompassing trans-media policy in order to bond the consumer to the cultural product and to the variety of its spin-off products. Such experiences with the branded product deepen young consumers' emotional bonding to the product, and partially shape their own identities (see chapter in this volume by Willett).

Even though Michel de Certeau (1988) has taught us that consumers are never passive, nor docile, children's play is partially shaped by commercial strategies as well as the definition of myths about childhood in contemporary society:

> Global circulation of the belief that 'real' childhood is organized by a combination of toys, fun, games, fantasy and controlled adventure, appropriately packaged for 'age' and 'stage', is essential to market expansion not in the sense that cultural construction precedes the market, but in the sense that this version of childhood is constituted through the market. The two are inseparable. (Langer, 2004, p. 256)

NOTES

1. Most of the information in this section is coming from SPFA (*Syndicat des Producteurs de Films d'Animation*), and especially the June 2006 edition of SPFA's publication *Le marché de l'animation*, a report published every year.
2. Usually, an animation series is sold to a channel that invests a lot, but asks for exclusive first broadcast; once the series has been totally aired for the first time, other channels, more often cable or satellite ones, are allowed to buy and air it; to secure that broadcast, some of them 'pre-buy' the programs at a very early stage.
3. A production is considered 'of French initiative' if the French producer has artistic control of the program (creating it or buying the rights linked to it) and is in charge of budget and fundraising.
4. Source: Commercial leaflet sent to licensees by Marathon 2007.
5. *Cat's Eyes*, a series about three girls who search for their father, was a huge Japanimation success in France in the 1980s, targeting 8- to 12-year-old girls.
6. *Cat's Eyes*, a series about three girls who search for their father, was a huge Japanimation success in France in the 1980s, targeting 8- to 12-year-old girls.
7. The 'design bible' is a document given to the licensees that imposes the rules for conceiving and adapting by-products inspired by the series, offering, most of the time, whole sets of adapted graphics.
8. The MIGUZI block was a famous TV block on Cartoon Network devoted to Japanimation.
9. Source: Commercial leaflet by Moonscoop 2007.
10. Source: Email received on March 19, 2005 at 22:36.
11. Source: A 17-year-old boy from the *Code Lyoko* community. Email received on April 29, 2005 at 16:53.
12. Declaration by Axel Dauchez, General Manager of Moonscoop, in *Ecran Total*, May 2005.
13. http://groups.msn.com/CodeLyokoFanGroup/frenchschools.msnw (Retrieved June 29, 2005.)

REFERENCES

Bakhtin M. (1968). *Rabelais and his world*. Cambridge, MA: M.I.T. Press. (Russian Edition: 1965)

Barthes R. (1974). *S/Z*, London: Cape.

Chandler D. (1995). *The act of writing: A media theory approach*. Aberystwyth: Prifysgol Cymru, UK: The University of Wales Press.

Consalvo, M. (2006). Console video games and global corporations: Creating a hybrid culture. *New Media and Society*, 8(1), 117–137.

de Certeau, M. (1988). *The practice of everyday life*. Berkeley, CA: University of California Press.

de la Ville V. I. (2007). The Consequences and Contradictions of Child and Teen Consumption in Contemporary Practice. In V.-I de la Ville (Guest Ed.), *Society and Business Review*, Special Issue, 2(1), 7–14.

Genette G. (1992). *Palimpsestes—La littérature au second degré*, Seuil, Coll. Points Essais, Paris.

Genette G. (1997). Palimpsests: literature in the second degree, translated by Channa Newman and Claude Doubinsky. Lincoln: London: University of Nebraska Press.

Jenkins H. (2006). *Convergence culture: Where old and new media collide.* New York: NYU Press.

Klein N. (2000). *No logo.* Toronto, Canada: Alfred A. Knopf.

Kline, S., Dyer-Witheford, N., & De Peuter, G. (2003). *Digital play—The interaction of technology, culture and marketing.* Montreal, Canada: McGill-Queen's University Press.

Kristeva J. (1980). *Desire in language: A semiotic approach to literature and art.* New York: Columbia University Press.

Langer B. (2004). The business of branded enchantment—Ambivalence and disjuncture in the global children's culture industry. In D. Cook, (Guest Ed.). *Journal of Consumer Culture, 4*(2), Special issue on Children's Consumer Cultures, 251–277.

Rifkin J. (2000*). The age of access: The new culture of hypercapitalism where all of life is a paid-for experience.* New York: Putnam.

Tobin J. (Ed.). (2004). Conclusion: The Rise and Fall of the Pokémon Empire, (pp. 257–292). In Tobin J. (Ed.) *Pikachu's global adventure—The rise and fall of Pokémon.* In Durham, NC: Duke University Press.

4 Consumption, Production and Online Identities
Amateur Spoofs on YouTube

Rebekah Willett

In April 2007, 45 million people visited the video sharing site YouTube, and each spent nearly 41 minutes there. Combining those minutes and converting them to years, these figures equate to 3,510 years spent watching YouTube in one month. Never before has people's play with moving image been viewed and discussed as it is at this moment. It is no surprise that YouTube has captured the attention of journalists who have written about 'clip culture' and the 'YouTube' effect, companies who have brokered deals with YouTube to set up channels and release copyrights on their videos, and of course Google, who bought YouTube for $1.65 billion in October 2006 (Geist, 2006). With the sale of YouTube, one of the questions being asked is how the site is going to start deriving revenue. Previously free from advertisements, how YouTube will be monetized is not yet clear.

However, the YouTube phenomenon concerns much more than time and money. For many people who have accounts on YouTube, the website offers ways of performing and defining identity. In addition to distributing videos (homemade or downloaded clips from elsewhere), YouTube account holders can display a selection of favourite videos, develop playlists, join groups dedicated to similar interests or styles of videos, display comments from other people, build a base of subscribers and subscribe to other YouTube accounts. All of these activities are visible on the homepage of the YouTube account holder, in addition to other personal information such as age, location, star sign and links to other websites such as MySpace. Furthermore, account holders are able to make personalized choices about the design of their homepage. Through the videos they post and through this information on their homepages, young people on YouTube can be seen as constructing their identities. However, their agency in performing their identity online is structured by many factors, not least the homepage templates provided by YouTube.

As part of the section in this volume which focuses on contexts of digital play, this chapter looks at structures which are framing young people's online productions. In some cases the structures are similar to the different types of inter-textual references described by de la Ville and Durup in their chapter. In other cases the structures could be examined in terms of the convergences between digital media cultures and marketing strategies, as Beavis does in her chapter. I am seeking to explore the relationships between the *structures* inherent in the media upon which young people are drawing and the *agency* (the capacity to think and act freely) of the young consumer/producer. My focus, then, is on the tension which underlies many debates about young people's online activities, between seeing young people as acted upon by media structures and societal forces and seeing them as independent actors in their own right. Instead of seeing structure and agency as a dichotomy, I will explore how structures on YouTube and the use of commercial media texts in young people's productions work to promote agency. The convergence here is about new technologies and commercial media texts and their relationship with young people's identities. The chapter starts by examining the role of commercial media more broadly in young people's identity work. Following this exploratory section, the chapter focuses on one example of young people's identity work in relation to media texts—the production and online distribution of amateur spoofs (or parodic sketches). The content and form of commercial and amateur spoofs are discussed, and theories of play are utilised to analyse the way the structures of media texts are used by young people in their performance of identity.

YOUNG PEOPLE, IDENTITY AND CONSUMER CULTURE

Sociological research has argued that the category 'youth' is a social and historical construct; yet at the same time, there has been research which examines how young people actively construct youth identities. Within studies of 'youth subcultures', analyses have focused on how material objects are used as markers of identity, defining specific social groups, and distinguishing them on the grounds of class, race and gender as well as age (Hebdige, 1979; Lury, 1996). Products from popular media are seen here as shared 'symbolic resources', providing easily accessible markers of interest and identity amongst young people. As society becomes increasingly fragmented by age, so too does the growth in products available to specific age groups. For example, Cook (2004) describes how the children's clothing industry worked to define particular subcategories of the children's market, through developing the 'toddler' and 'teen girl' categories of clothing. However, this is not to say that markets *create* different categories of childhood. As Cook argues, 'they provide, rather, indispensable and unavoidable means by which class specific, historically situated childhoods are made material and tangible' (p. 144).

Ideas about the relationship between consumer items and identity apply equally well to online cultures. For example, social-networking sites which combine blogs, profiles and photo and video-sharing can be viewed as cultural resources which are used by young people as a way of performing and perhaps playing with their identity. These sites often contain references to consumer culture—for example, personal web pages often feature the author's favourite music which plays when a user accesses the page. Furthermore, commercial websites offer children and young people specific identities connected with consumer culture. Websites targeted at tween girls, for example, reflect a particular market discourse which attempts to capitalise on the emergence of the category 'tween' (Willett, 2005). Referring to the dual nature of the audience and marketing culture, Quart (2003) describes how consumer culture not only brands teens as subjects, but also positions teens as branded objects. However, young people are not simply passive victims of this process: on the contrary, consumer culture increasingly positions them as active participants within it.

It is important to recognise that young people do not necessarily consume an item 'straight off the shelf'. For example, McRobbie (1991) discusses how with girls, personal style becomes a focus for display, particularly as they grow older and interact independently in more public spaces (away from shopping trips with mum, for example). Furthermore, using the example of how girls use second hand clothing, McRobbie argues that girls resist, choose, rework and recreate consumer trends. Several researchers in this field use Levi-Strauss's (1974) notion of 'bricolage' to describe how young people draw on a variety of sources and then piece together, recontextualise and transform cultural items to create a new self-image or identity. Homepages, for example, are analysed by Chandler and Roberts-Young (1998) in terms of 'bricolage', referring to the processes involved in creating a page made up of references and images from various sources which have been appropriated and recontextualised. In including, omitting, adapting and arranging these references, the 'bricoleur' is also constructing and performing an identity. Viewing consumption in these terms, we can see young people as active agents, appropriating consumer culture for their own uses.

For theorists such as de Certeau (1984), ordinary readers of cultural products write their own meaning by selecting and transforming meanings of cultural commodities in response to the 'strategies' of dominant institutions. The practice of 'poaching' texts, using de Certeau's term, is seen by some as an important aspect of active consumption, one which is leading to new participatory media cultures. Jenkins (2006) looks specifically at the activities of fans as an example of a participatory media culture, and argues that the convergence between fans and industries is creating a new landscape. Jenkins describes instances in which fan productions, ideas and ways of communicating and sharing knowledge are recognised by media industries and incorporated into their products or their way of developing and marketing a product. These fan practices attest to the power of the consumer to challenge industries and to force new relationships between consumers and producers.

One way in which fans create new meanings in their productions is through elements such as parody, providing a critical take on texts or cultural practices. Applying Jenkins's approach, parody can be seen as more than just play for the sake of playing—parody is about redistributing power from the producer, allowing the consumer to contest, critique and create their own media, and do so in a way which has an impact on big businesses. In contrast, theorists such as Jameson (1984) argue that postmodernism is characterised by a culture of pastiche or 'blank parody', consisting of imitations of previous styles with no critical edge or ulterior motive. In Jameson's view, postmodern society is void of creativity, constantly rehashing the same ideas, images and quotations. Jameson describes postmodern culture as consisting 'of flatness or depthlessness, a new kind of superficiality in the most literal sense' (1984, p. 60). Jameson's ideas raise questions about the intent of authors and the content of parodic texts as well as the extent to which consumers as producers, such as those described by Jenkins, are changing industry practices on any significant scale. In these debates structure and agency are formulated in different ways, with Jameson attributing little agency to modern audiences who are merely replicating existing structures, and Jenkins portraying an optimistic view of audiences as challenging and changing structures within texts, industries and participatory cultures. The following sections engage with these debates by analysing how the structures of media texts support young people's participatory media practices.

THE STUDY

My aim here is to examine how the content and form from commercial structures are used by young people in their spoofing practices and explore what else might be happening in these practices. The study is part of a larger project which looks at 'camcorder cultures' in the United Kingdom.[1] Spoofs were chosen as a style of production after initially surveying over two hundred videos on YouTube and categorising them according to content and purpose. The terms spoof and parody are used interchangeably on YouTube, although spoof is used far more frequently. Spoofs are a common practice in commercial texts (*The Simpsons*, e.g., is a parodic depiction of a suburban family who regularly watch a spoof of *Tom and Jerry*). In amateur contexts, spoofs range from productions which imitate a text in a playful way, through to satire in which the author clearly intends to problematise and criticise a text or practice.

With tens of thousands of spoofs at one's fingertips, narrowing the field down to make a manageable research project is necessary. Using Google's search engine, I located specific types of spoofs on sites such as YouTube and Google video, and in the end located over 120 spoofs from 68 different producers with the terms I identified. I did not attempt to get a representative sample of spoofs on video sharing sites, and the sample is particularly limited in that most of the spoofs were labelled with the search terms I specified. Therefore I am not including many other texts which might also fit the definition of a spoof. Given the aim of wider project, to look at amateur camcorder use in the United Kingdom, of which this study is one part, I searched for spoofs with original footage that were clearly made by amateurs living in the United Kingdom. I searched with the keywords spoof, parody, satire, homage, pastiche and mockumentary (individually). 'Spoof' brought up the most results by far, and featured in the title, tag or description of videos. 'Parody', 'satire', 'homage' and 'mockumentary' brought up a smaller number of results, and 'pastiche' did not lead to any results within my defined area. The producers qualify their videos in the description, tags and sometimes title, making it clear that the video is a spoof and not a serious imitation of a text. These qualifying statements include labels such as fake, joke, comedy, funny, humor/humour, crazy, weird, stupid and silly. Furthermore, a number of producers make it clear that their posts are projects they did when they were bored or had nothing better to do. (This way in which authors distance themselves from their product is worth investigating, but I do not have the space to do so here.)

A large majority of the producers of the spoofs surveyed are young white men, aged approximately 12–25, and we can assume they have access to camcorders, editing equipment and broadband internet access. These profiles are confirmed by information given on YouTube accounts which match the personal appearances in numerous videos on each account and information gleaned from linked websites, such as MySpace pages. (Only 3 out of the 68 producers were women, though occasionally women were included as actors. Only 4 out of the 68 were over the age of 25, and all were under 40 years old.) After conducting this widespread survey, I selected four case studies to look at more closely. The data for these case studies were collected through phone or email interviews and through the comments left by viewers on the video sharing sites. The case studies were chosen on the basis that they were U.K. amateur producers, clearly identifying their work as spoof and having feedback on their videos which could be accessed through the comment sections online. The cases typify a majority of the 120 spoofs surveyed, being produced by groups of friends, with young white men predominating. The following sections include an analysis of the data collected from the broad survey of spoofs as well as an analysis of one of the case studies.

SURVEYING ONLINE SPOOFS

As mentioned previously in relation to *The Simpsons*, spoofs are a common form of entertainment in a media consumer's diet: comedy sketches which spoof other programmes or cultural practices; political or news-related spoofs; advertisements which spoof other advertisements, music or cultural practices; and movies which parody entire genres (disaster movies, for example). Spoofs also make the news, particularly when political lines are crossed (e.g., VW Polo's spoof advertisement featuring a suicide bomber, or the BBC's spoof of Al-Jezeera news). The form and content of spoofs online, therefore, reflect these examples from media industries. Amateur spoofs include comedy sketches, newscasts, music videos, advertisements and horror movies. The producers I interviewed also commented that they get their ideas from current and older sketch shows such as *Monty Python's Flying Circus*. Content common to amateur spoofs such as sexual references, language/accent, cultural references (e.g., class-based stereotypes of clothes) are exemplified in professional spoofs, particularly sketch shows such as *The Catherine Tate Show* or *Little Britain* in the United Kingdom or *Saturday Night Live (SNL)* in the United States.

Professional spoofs often play with content—providing examples for the amateur spoof producer. An advertisement for a high definition television (Sony Bravia) with the strapline 'colour like no other', was spoofed by the drinks company Tango, who changed the setting from luxurious San Fransisco to a small rundown town in the United Kingdom, light bouncing balls to whole fruits which smash and destroy property and the strapline to 'refreshment like no other'. In turn various spoofs were produced by amateurs, included a machinima spoof (machine animation produced on video game platforms) which featured soldiers hopping down a hill and the strapline 'bunnyhopping like no other'.[2] Advertisements provide ready-made structures for spoofs and are popular forms for amateurs. In addition to providing a narrative and music, the Sony Bravia ad contains particular shots that are integrated into spoofs (e.g., panoramic views of a city, a shot of a street from an alley). The structure gives the amateur a way of signalling familiar cultural objects to potential audiences, and because the structure is familiar, the focus is shifted to the content. The content, therefore, becomes a key feature of an amateur spoof (although online discussions often include the mechanics of obtaining particular imitative shots, music, locations, etc). With the focus on content, spoofs can signal a range of purposes, from serious critique through to playful imitation. In the case of the bunny hopping soldiers, it is safe to assume that the authors were not critiquing Sony Bravia or the videogame. The spoof was more a playful imitation—but at the same time the humour derives from the cultural understanding that bunny hopping is the antithesis of a soldier's mode of movement, and therefore in having a group of bunny hopping soldiers, the author highlights the stereotype of power, stealth and heavy marching associated with soldiers.

In some cases, it is clear that the form has a crucial impact on the content of the spoof. Instead of aiming to spoof a particular cultural phenomenon, for example, the form determines how something is going to be filmed and discussed. A holiday film of a family in a seaside caravan, for example, is shot as a spoof of *Big Brother*, with snippets of the mundane daily activities and each family member coming to 'the boiler room' to talk about their relations with other members of the family.[3] By spoofing *Big Brother*, the form is lending itself to a commentary and a particular way of commenting on the holiday activities. By using the form of *Big Brother*, one could argue that the text holds up the form for critique—forcing someone to sit in front of a camera (in the diary room) and interrogating them about their actions, attitudes and feelings; getting people to comment on their relationships in private and then broadcasting those comments publicly; nonstop filming of everyday interactions. These are forms that one might critique. However, the spoof is also saying something about a caravan holiday—it is like the situation of being stuck in a house with very little to do except analyse one's relationships with each other, and with someone creating activities for everyone to take part in. Therefore, the video could be read as a parody of the form and content as well as the production itself.

Another way of looking at parodic media productions is as a practice of play. Silverstone (1999) connects play with media consumption, and uses theories of play to discuss broader practices of cultural production. He writes: 'In play we investigate culture, but we also create it' (Silverstone, 1999: 64). Spoofs involve playing with form (e.g., documentary, advertisement, reality TV show) as well as content (e.g., representations of gender, age or class). For example, a spoof day-in-the-life documentary of a particular culture can be seen as a way of investigating cultural stereotypes as well as the practice of documentary form. Further, by holding up a stereotype for ridicule, a practice is created amongst the group producing the spoof as well as the wider audience who watch the spoof.

Fiske's work on children's television culture discusses play with television texts in terms of the rules established by the texts (Fiske, 1989). By playing with a text (e.g., creating a playground game or a spoof based on the text), the player is necessarily following particular rules (content or form, as discussed above). By following the rules, as in a game, one is allowed to participate, to experience pleasure in the game. According to Fiske: 'The pleasures of play derive directly from the players' ability to exert control over rules, roles and representations' (p. 236). Importantly, as Fiske asserts, participation then allows freedom to question the rules, to explore meanings and identities offered through the game. Through play with texts, roles and representations are chosen, and can be replicated but equally they can be inverted. This interplay between control or *ludus* (following rules) and freedom or *paidia* is discussed throughout play theory. Rather than seeing play as a cultural practice (Huizinga, 1938), play can be seen as a system of interactions between reality, fantasy, roles, rules, subjects and objects (Caillois, 1961). This is particularly important when considering questions of structure and agency in relation to media consumption and production. Spoofs offer a way of repurposing media and cultural texts, a way of playing with the structures and using them for different kinds of interactions—for example, exploring masculinity, friendships, fantasies or positions of power.

Clearly the multiplicity of meanings in the text need further reference points in order to understand their consumption and production. The next section, therefore, looks more closely at one group producing online spoofs, analysing one of their productions and examining the convergence between structures of the text from which they are drawing, the affordances of YouTube and the identity work of the young producers.

THE BENTLEY BROS

As mentioned previously, one of the arguments about participatory media is that it is allowing amateur producers to create new meanings from existing media texts, develop different levels of communication and forge new relationships with the media industry (Jenkins, 2006). The case I discuss here provides possible evidence for these arguments, but I also suggest that, at least in reference to the study I have done, there are other more local interactions occurring that are worth examining. Undeniably, on one level there are a greater variety and number of amateur productions being distributed, viewed and discussed. There are wider implications for this—so, for example, one of the four young men I interviewed for this study of spoofs said: 'I am finding that a lot of people are talking about what they saw or found on YouTube last night as opposed to a year or two ago when people would chat about what was on TV the night previously'. On another level, however, these amateur productions can be seen as more localised—the productions I looked at could be described as being more about groups of friends playing with a piece of technology and 'having a laugh'. Here the production can be seen a conduit for defining and performing identities, creating different meanings from existing texts and critiquing or 'holding a mirror' to cultural practices.

The case study I focus on here comes from a group of four brothers and their friends, ages 14–20, called the Bentley Bros. The group live in a small town in rural England and describe themselves on YouTube as 'a small group of people, that make homemade comedy films with nothing more then a cheap camera and editing software'.[4] According to Stuart Bentley, all of their 15 productions online are comedy, with the exception of one experimental horror film. The comedy often centres on spoofs of thriller-type genres, and includes direct references to TV, videogames and movies, including *Resident Evil*, *Ghost Hunters*, *The Babysitter* and *Scream*. The productions vary from carefully scripted and directed 60 minute films, complete with rented costumes and props which take over a month of intense work, through to short films shot in an evening which are improvised on the spot.

The group has a strong fan base, as can be seen on the Bentley Bros own website[5] and on YouTube.[6] At the time of writing this chapter, their website has a section of fan art, links to four fan sites, and a forum with over 500 registered users and over 50,000 posts. On YouTube, their production 'Resident Evil 4' has been viewed 270,000 times, received over 1,500 comments, and at one point was number 56 in the all time top rated videos (games and gadgets category). One of the questions to ask is whether groups like the Bentley Bros are part of Jenkins's 'convergence culture'. On the one hand, we could argue that the Bentley Bros' productions do not fit with a new model of media production. The Bentley Bros' productions have only been distributed online, and one could argue that the industry has not incorporated or acknowledged amateur productions such as this to any significant degree. Furthermore, although the Bentley Bros have a large fan base, the productions are still restricted to a small group. In fact, one of the frustrations the Bentley Bros express is in finding actors and new locations for filming. In terms of sharing knowledge, Stuart Bentley commented: 'We never really learnt our editing/filming from anyone nor got advice except from the films we watch. But you need the knack to notice the work behind a film rather than get drawn into it'. None of the producers I interviewed had any formal training on filmmaking, and they all said they picked it up from using the equipment and through watching movies and other media. The question these responses raise, therefore, may not be about converging culture and new forms of knowledge sharing, but may be more about media literacy. What is the relation between consuming and producing media? How are common media technologies such as camcorders and simple editing software scaffolding learning?

On the other hand, one could argue that the productions are part of a wider trend which is changing the industry. Jenkins (2006) argues that fan productions previously ignored by the industry are now being noticed:

> No longer home movies, these films are public movies—public in that, from the start, they are intended for audiences beyond the filmmaker's immediate circle of friends and acquaintances; public in their content, which involves the reworking of popular mythologies; and public in their dialogue with the commercial cinema. (p. 143)

This has meant that industries have felt the loss of control of their materials—Jenkins gives examples of copyright infringement policies, including Warner Bros.' actions to close down *Harry Potter* fan sites, many of which are created by children and young people. However, Jenkins argues that industries also adopt fan practices—looking at experimental practices online that have attracted cult followings and trying them out themselves. The challenge for industries, Jenkins argues, is to make distinctions: 'between commercial competition and amateur appropriation, between for-profit use and the barter economy of the Web, between creative repurposing and piracy' (p. 167). It is difficult to measure the impact of one group, such as the Bentley Bros, on industry practices. One could argue that they are part of the larger group of fans who are remaking texts and developing substantial audiences, and Jenkins provides evidence that the industry is taking account of these groups.

Certainly the Bentley Bros have a much larger audience than they would have before the ease of videosharing. The audience goes far beyond the small village where the Bentley Bros live, and from the comments it is clear some of the audience members are outside the United Kingdom. Although the audiences are not giving the Bentley Bros advice or sharing formal kinds of knowledge, they are providing feedback on their films and adding to a general discussion about media cultures. In this way, Jenkins's discussion of the public nature of films and the reworking of media is applicable. In order to examine more closely how this reworking of media is happening, the next section focuses on one of the Bentley Bros' films entitled 'Bug Busters'.

'BUG BUSTERS'

Although there is a film from 1998 of the same title, the Bentley Bros' 'Bug Busters' is a general spoof of action/adventure/shoot 'em up movies with little relation to the 1998 film. The description on YouTube reads: '4 Men . . . One Mission. To rid the Earth of Bug Scum'. The film starts with four action heroes in their home, three doing physical exercises and one reading a book. The heroes are dressed vaguely as 'ninjas', all in black with black bands around their heads, and one has two swords attached to his belt. The sound track at this point is a phone conversation, with a young man requesting the service of exterminators for 'a minor bug problem', and a low gravelly voice insisting that 'no bug problem is minor . . . these things are killers, all of them, and they must be stopped'. After the title sequence, the heroes are next seen in a line walking toward the camera, all in black, heavily laden with weapons. Special effects have been added so that the heroes come into focus as they walk toward the camera, and strong rhythmic music announces their entrance at the door of a house. The movie proceeds with a carnage of flies and destruction of the house, as the Bug Busters shoot flies with various guns, slash them to pieces with swords and eventually blow up the house. The final shot is of the heroes about to kill a very large spider outside the house with an equally large sling-shot.

It is possible to do a textual analysis of the film and discuss various references to other action/adventure films, including music, dialogue, camera angles, and particular moves by the actors. This is one of the things that audiences frequently do in the comments to the Bentley Bros' films. In making these inter-textual connections, the Bentley Bros are signalling to their audiences what their interests are and what they expect their audiences to know. In this way, they are defining their identities as viewers of particular films. Furthermore, the Bentley Bros are repurposing these references by putting different elements together (parodic convention of men dressed up as women alongside men with big guns) and holding up particular conventions as parody (shoot-em up scenes). In doing so, the Bentley Bros are inviting different readings of media texts—the comic side of using oversized and multiple weapons in shoot 'em up scenes, for example. One of the interesting elements of parody is the ambiguity of the text. Are the Bentley Bros making fun of action/adventure films, ninja films, one particular film (*Ghost Busters* or *Bug Busters*), audiences who take action/adventure seriously, the representation of masculinity in action/adventure, or even themselves as young men pretending to be superheroes? Are they holding these things up for critique or are they simply representing them in a playful way?

If we examine 'Bug Busters' as a space in which the Bentley Bros are playing, we see further ambiguity. Bateson argues that play is always twice removed from reality. In his words, 'The playful nip denotes the bite, but it does not denote what would be denoted by the bite' (1972, p. 180). The brandishing of the swords in 'Bug Busters' is not denoting the anger or fear represented in the text from which they might be drawing (any number of ninja movies, for example) nor the original sources from which those movies draw (Japanese assassins active from the 15th–19th centuries). In this sense, play is a fantasy space which allows for the exploration of meanings and roles found in texts. It is in this play space that discussions about creativity arise. Media theorists have used Winnicott's ideas to discuss fan and gaming practices as important spaces for creative expression, for exploration of meanings of cultural texts and for 'psychic health' (Dovey & Kennedy, 2006; Hills, 2002). Winnicott refers to the psychological dimension of defining oneself in relation to others, particularly in the transition from childhood through to adulthood. For Winnicott, successful transition is marked by creative acts of communication and use of objects, such as the use of an object to symbolize a relationship or a feeling. Play provides a transitional or third space in which the subject engages in communications and creative acts which mark transition and the definition of the self. Here the experience of inner and outer worlds, of fantasy and reality overlap, and creative involvement with subjects and objects takes place. 'It is in playing and only in playing that the individual or adult is able to be creative and to use the whole personality, and it is only in being creative that the individual discovers the self' (Winnicott, 1971, p. 54). Using Winnicott's theories and viewing the Bentley Bros' movies as forms of play, the productions can be seen as providing important spaces in which the young men are exploring the meanings of media texts in relation to their own identities.

Importantly, it is in this third space that the meaning of cultural objects is negotiated and in which dominant discourses can be contested. This negotiation of meaning overlaps with Bhabha's ideas about third space. Bhabha (1994) focuses on discourses and the resulting hybridity that occurs within cultural signs or symbols when people and ideas come together. In Bhabha's notion of third space, different cultural signs and competing discourses exist alongside each other rather than eliding into each other, and in doing so new meanings are able to be produced. Although both Winnicott and Bhabha discuss the potential for third spaces to challenge dominant discourses, there is equally the possibility to accept and replicate discourses. Concerns over children playing with heavily commercialized toys (Disney princess characters, e.g.) reflect this view of play as potentially reinforcing rather than challenging dominant discourses. Similarly, one could argue that the Bentley Bros' play with dominant discourses around masculinity are replicating those discourses rather than producing new meaning through them.

However, in the reality side of Winnicott's 'third space', the boys are putting themselves in powerful positions of director and producer. The production as a play space allows the group of friends to express their friendship, their media savvyness, their identity as achievers and their skill in using technology. From this viewpoint, the production process offers an important way for the friends to define themselves in terms of discourses around young men. By producing a video which demonstrates technical skill and draws on popular culture, as well as marking the video as a spoof (not serious), the Bentley Bros are negotiating their position as serious achievers and 'cool' boys (for a discussion of the difficulty of negotiating this position, see chapters by Jackson & Epstein in Epstein, Elwood, Hey, & Maw, 1998). The overwhelmingly positive response to their productions that is displayed online demonstrates their success at this negotiation and provides them with further incentive to continue performing their identity in this way.

One final aspect of media productions to consider is in relation to their identity as friends. Media are part of the social life of young people, and through discussing and producing media, friends are not only connecting on a social level, they are also making sense of media in their lives (De Block & Buckingham, 2007). Furthermore, humour, an essential aspect of spoofs, is a common element of 'mateship culture' as analysed by James and Saville-Smith (1989). Kehily and Nayak (1997), who look specifically at male humour in secondary schools, conclude, 'humour plays a significant part in consolidating peer group cultures . . . offering a sphere for conveying masculine identities' (p. 67). Producing a spoof of an existing media text offers young men the chance to display their friendship and their masculinity. However, the productions can also raise questions about masculinity. As 'Bug Busters' demonstrates, dominant masculinity is parodied, and many spoofs feature men dressed up as women. Soep's (2005) study of a group of young men making a film in their basement shows how off-camera interactions are contrary to the 'hard-core masculinity' as displayed in their production. Soep writes: 'their moment-to-moment interactions revealed a very different ideology—one coded in everyday life and scholarships as "feminine", characterized by intimacy, propriety and intense collaboration'

(p. 178). The parody of media texts for the Bentley Bros, therefore, works to cement their relationships, express their identity, give them a common point of reference, but also a way of processing media—making sense of representations of masculinity, 'violence' and horror in films, for example. Importantly, it is through play with media texts and through new forms of participatory media that these interactions are occurring.

CONCLUSION

Young people's identities are being performed and defined in commodified environments such as YouTube, and they are using commercial media in their video productions to make statements about themselves. I have given the example of structures in spoofs which are supporting young people's participation as media producers. Here the content and form from commercial spoofs are providing examples which young people draw on in their own productions. Furthermore, the texts which are being spoofed provide content and forms that young producers are using to make new meanings from media texts and surrounding practices. Finally, sites such as YouTube are providing new means of distribution, expanded ways of displaying identity online and the possibility for large global audience interaction. In these various ways, I argue that young people's agency works through the structures inherent in new media participation such as online spoofs. New technologies and commercial media texts are structuring young people's identity work, but they are also providing the resources through which young people are exploring identity issues, critiquing media texts and examining cultural practices.

NOTES

1. The project 'Camcorder Cultures: Media Technologies and Everyday Creativity' is funded by the U.K. Arts and Humanities Research Council (reference number RG/112277) and based at the Centre for the Study of Children, Youth and Media, Institute of Education, University of London.
2. www.youtube.com/watch?v=s0FLSr4S9rQ, Retrieved July 3, 2007
3. www.youtube.com/watch?v=TrMKYwmAwBY, Retrieved July 3, 2007
4. www.youtube.com/bentleybros, Retrieved July 3, 2007
5. www.bentleybrosproductions.com, Retrieved July 3, 2007
6. www.youtube.com/bentleybros, Retrieved July 3, 2007

REFERENCES

Bateson, G. (1972). *Steps to an ecology of mind: Collected essays in anthropology, psychiatry, evolution and epistemology.* London: Granada Publishing.
Bhabha, H. (1994). *The location of culture.* London: Routledge.

Caillois, R. (1961). *Man, play and games*. New York: Free Press.
Chandler, D., & Roberts-Young, D. (1998). The construction of identity in the personal homepages of adolescents. Retrieved July 3, 2007, from http://www.aber.ac.uk/media/Documents/short/strasbourg.html
Cook, D. (2004). *The commodification of dhildhood: The children's clothing industry and the rise of the child consumer*. Durham, NC: Duke University Press.
de Block, L., & Buckingham, D. (2007). *Global children, global media: Migration, media and childhood*. Basingstoke: Palgrave.
de Certeau, M. (1984). *The practice of everyday life*. Berkeley, CA: University of California Press.
Dovey, J., & Kennedy, H. (2006). *Game cultures*. Buckingham, UK: Open University Press.
Epstein, D., Elwood, J., Hey, V., & Maw, J. (Eds.). (1998). *Failing boys? Issues in gender and achievement*. Buckingham, UK: Open University Press.
Fiske, J. (1989). *Understanding popular culture*. London: Routledge.
Geist, M. (2006, March 20). The rise of clip culture online. *BBCNews*. Retrieved July 3, 2007, from http://news.bbc.co.uk/1/hi/technology/4825140.stm
Hebdige, D. (1979). *Subculture: The meaning of style*. London: Methuen.
Hills, M. (2002). *Fan cultures*. London: Routledge.
Huizinga, J. (1938). *Homo ludens. Versuch einer Bestimmung des Spielelements in der Kultur.* (*Homo Ludens—A Study of Play Element in Culture.*) Hamburg, Germany: Rowohlt.
James, B., & Saville-Smith, K. (1989). *Gender culture and power: Challenging New Zealand's gendered culture*. Aukland, New Zealand: OUP Australia and New Zealand.
Jameson, F. (1984, July–August). Postmodernism, or the cultural logic of late capitalism. *New Left Review, 146*, 53–92.
Jenkins, H. (2006). *Convergence culture: Where old and new media collide*. New York: New York University Press.
Kehily, M. J., & Nayak, A. (1997). 'Lads and laughter': Humour and the production of heterosexual hierarchies. *Gender and Education, 9*(1), 69–87.
Lévi-Strauss, C. (1974). *The savage mind*. London: Weidenfeld and Nicolson.
Lury, C. (1996). *Consumer culture*. Cambridge, UK: Polity Press.
McRobbie, A. (1991). *Feminism and youth culture: From 'Jackie' to 'Just Seventeen'*. Basingstoke, UK: Palgrave McMillan.
Quart, A. (2003). *Branded: The buying and selling of teenagers*. London: Arrow Books.
Silverstone, R. (1999). Rhetoric, play performance: Revisiting a study of the making of a BBC documentary. In G. Jostein (Ed.), *Television and common knowledge* (pp. 71–90). London: Routledge.
Soep, E. (2005). Making hard-core masculinity: Teenage boys playing house. In E. Soep & S. Maira (Eds.), *Youthscapes: The popular, the national, the global* (pp. 173–191). Philadelphia: University of Pennsylvania Press.
Willett, R. (2005). Constructing the digital tween: Market forces, adult concerns and girls' interests. In C. Mitchell & J. Reid-Walsh (Eds.), *Seven going on seventeen: Tween culture in girlhood studies* (pp. 278–293). Oxford, UK: Peter Lang.
Winnicott, D. W. (1971). *Playing and reality*. London: Tavistock Publications.

Part II

Children and Digital Cultures

Introduction to Part II

In this part of the book we move to a close examination of actual behaviours during the processes of consumption, through detailed ethnographic studies of children's engagement with digital cultures, covering a variety of media and contexts. As we see the carefully captured realities of these engagements, we can begin to understand just how complex and rich they are and how oversimplistic developmental models of childhood fail to explain adequately the patterns of children's behaviours in a digital age.

Clare Dowdall considers how one child performs and negotiates her social identities as she works and plays in a range of digital contexts in and out of school, using texts as key platforms for the rehearsal, performance and negotiation of social identities which contribute to the wider discourse of home, school and friendship. While most texts from these domains can be described as purposeful and socially-motivated, they are created in diverse social, cultural and physical spaces and are circumscribed in different ways. In particular, text production in some digital contexts invites collaboration between the performer and the audience, as feedback and comments are sought and posted. This reciprocity impacts upon the possibilities for the representation of self through text production and unsettles the notion of the digital text as a platform for agentive performance of social identity.

In today's world, commercial, technological and cultural pressures combine to create a climate where stories are much less likely to come in singular packages; instead they are adapted into different media, retold and resold. In Chapter 6, Margaret Mackey presents examples from longitudinal studies with children and adults as they revisit familiar texts and move between different media versions of stories, arguing that we learn how stories are created and interpreted from multiple incarnations of a particular fictional world. She draws on Peter Lunenfeld's discussion of contemporary culture in terms of its aesthetic of 'unfinish' and Kristie Fleckenstein's notion of 'slippery texts' which involve a variety of verbal and visual technologies to develop related concepts of thick play and big worlds. These then serve as organizers to explore how learning takes place in a world where many stories never seem to come to an end.

Similarly, in Chapter 7 Julia Davies draws on her ethnographic online research which spans across a number of years to describe some of the ways in which individuals use the Internet to play out a range of identities, to experiment with ideas and to make connections with others. Using data drawn from blogs, from an online photo-sharing community, and from discussions on online web boards, she shows how individuals create and use online spaces in ways that link to other spaces they inhabit, weaving complex multimodal discourse webs and patterns creating a range of narratives that cross boundaries and allow the development of new ideas. Her argument is framed in the context of new literacy studies and multiliteracies and multimodalities.

In the final chapter in this section we return to close observations of children, this time playing both online and in real physical space. In her analysis of these observations, Beth Cross draws on recent anthropological work which has highlighted the role of mimesis within play, a process whereby roles and activities are not only faithfully imitated, but are mimicked with an intentional twist. She suggests that integral to the creativity of play are 'repertoires of resistance' that take advantage of material and linguistic resources to spin or accentuate desired meanings, in differing ways, in different contexts. Ethnographic and discourse analysis material is presented that follows children's play from on-screen participation to off-screen recombination and transposition of digital motifs, scenarios and tropes in informal role playing activity. The chapter examines what children borrow from digital genres, and the ways they negotiate its use with each other in order to understand how these acts of negotiated mimesis constitute performances of social identities and social aspirations.

5 The Texts of Me and the Texts of Us
Improvisation and Polished Performance in Social Networking Sites

Clare Dowdall

> . . . *I suffer from my own multiplicity. Two or three images would have been enough, or four, or five. That would have allowed for a firm idea: This is she. As it is, I'm watery, I ripple, from moment to moment. I dissolve into my other selves. Turn the page: you, looking, are newly confused. You know me too well to know me. Or not too well: too much.*
>
> (Atwood, 2006, p. 25)

In this extract from her short story, *No More Photos*, Margaret Atwood describes how photographs, placed within an album, project various representations of who she is for her audience. The viewer of the album is described as being confused by these contrasting images, while Atwood recognises her own multiplicity. Atwood's short story neatly illustrates that the textual artefacts we create to represent ourselves to others can realise multiple social identities, which, when placed alongside each other, offer up an array of possibilities for the meaning maker. Traditional photograph albums have always had this potential. Now, online digital texts offer this possibility too.

Online digital texts are a feature and by-product of the early 21st century communication landscape. This contemporary landscape is subject to the rapid and insistent evolution of a variety of digital communication technologies (Carrington, 2005a, 2005b; Kress, 2005; Livingstone & Bober, 2005) that have facilitated a range of new communicative practices (Marsh, 2005). These acts of communication are observable in a variety of material and online digital spaces as people go about their everyday lives: the supermarket is occupied by shoppers using their mobile phones to check the details of their shopping list; mobile phones vibrate in classrooms and conference halls alike to alert someone that a new text message has been received; and in the privacy of their own homes, children continue the day's activities and events using MSN Messenger[1] and a host of social networking sites to maintain their friendships and relationships. These new communicative practices continue to evolve. They depend on access to mobile phones and the internet; access which, according to data published by the *UK Children Go Online* (Livingstone & Bober, 2005)[2] and the Office for National Statistics (2007),[3] is growing rapidly, thereby enabling more and more children and their families to participate.

This chapter will consider how one young adolescent, Clare, uses a specific type of online communicative practice to create textual artefacts which represent her to others within the early 21st century communication landscape. Clare is 14 and has been an avid user of the social networking sites Bebo[4] and MySpace[5] for the past 3 years. During this time, her use of the sites and sophistication as a user has evolved, along with the possibilities offered by the sites themselves. Clare is a participant in my ongoing doctoral research. She has been interviewed informally about her creation of texts for online social networking during the last 3 years and is part of a purposive sample who are being selected because of their avid use of online social networking and in particular, their creation of texts to represent themselves to others in online spaces.

To contextualise this discussion, an introduction that describes the profile pages and some of the features of Bebo and MySpace is necessary. dana boyd (2007) has recently overviewed these online digital contexts in some detail in her work for the MacArthur Foundation. Her article provides a comprehensive introduction that will help to orientate less familiar readers[6]: For the sake of coherence, I will attempt a brief overview here. MySpace and Bebo are social networking sites that can be joined by anyone with online access. Membership is free. To join, members have to create an account by submitting an email address and password. They can then create a profile page which acts as a space where photos, videos, music, lists of likes and dislikes and other personal information can be shared. Members email their friends to inform them that they have joined, and to invite their friends to join too as members with access to their profile page. Members can then view and comment on each others' pages in response to photos or interactive features. In this way an online social network is created. MySpace and Bebo continue to grow. According to the web information company Alexa (2007),[7] MySpace and Bebo are currently the 8th and 11th most visited websites in the United Kingdom, with MySpace being reported as having over 100 million members worldwide. To say that these social network sites have had a prolific impact on young people's online and offline activities does not really conjure the extent to which these sites pervade popular youth culture. The BBC has repeatedly reported on the phenomenal uptake of MySpace and Bebo in the UK since their launch.[8] boyd (2007) draws upon her ethnographic research in the United States with the social network sites MySpace and Friendster to support this claim.

As a regular user of Bebo and MySpace, Clare has created and continues to create profile pages to represent herself to her online social network as part of her everyday communicative practice. These pages are sophisticated interactive multimodal texts, containing words, still images, sound and video. Examples of her pages are included later in this chapter. Online profile pages are a far cry from the traditional photograph album that Atwood describes in her short story, but they serve to act as texts that capture and represent Clare to her chosen online audience. As such, these texts can be seen as a 21st century version of the predigital photograph albums that rendered identities and events in a fixed form. However, the affordances, materiality and ownership of these texts differ significantly from the predigital photograph album. Where a photograph album can provide a fixed artefact or an archive of a person's existence that can contribute to the positioning of the individual over time in society, an online profile page offers a more transient opportunity for identities to be rehearsed, improvised and performed. Bauman (2005) describes how 'liquid life' is lived in a 'liquid modern society': one in which the members of society change how they act before their actions consolidate into habits. As Clare continually evolves her online profile pages as part of friendship and social relationship work and play, her artefacts of social identity can be viewed as liquid rather than fixed. In this chapter, the potential for representing self to others using online profile pages will be considered, and in particular, the opportunities to rehearse, improvise and perform social identity within these spaces as online digital texts are created will be discussed.

My interest in digital text creation builds from two pieces of preliminary postgraduate research. First, through a study of four children's out-of-school text production, I found that some children produced texts in order to affiliate themselves with distinct peer groups such as 'texters' and computer 'gamers' (Dowdall, 2006a). Building from this finding, preliminary ethnographic research with Clare, a then preteenage child, introduced me to the world of online social networking and some social network sites that are used prolifically by young people, for example, Bebo and Myspace. Through this work, it became apparent that Clare orchestrates a range of masteries, including a social mastery, in order to negotiate friendships and produce texts in online contexts (Dowdall, 2006b). In tandem with the wide acceptance of the New Literacy Studies paradigm (Marsh, 2005), where literacy is regarded as a social practice (New London Group, 1996; Barton & Hamilton, 2000), the use of the internet as a site for constructing and playing with social identities has been the subject of much recent academic interest. Based on ethnographic research that was conducted in the late 1990s, Thomas (2004) identified that girls use words and images in cyberspace to 'write' cyberbodies as they consciously construct their desired identities for their audience. This early view of singular and bracketed identity-play, where a subject creates an avatar in an online context to represent self, has evolved as the affordances of digital technologies have expanded. Through exploring seven teenagers' use of instant messaging,

Lewis and Fabos (2005) have identified 'multivoiced' social subjects who assume alternative social identities as they navigate their relationships in online spaces. This view suggests a more fluid use of digital technologies for online social identity play, and a context that is more closely related to the everyday realities of the subject involved. Leander and McKim (2003) develop the idea that online and offline social identity work and worlds are interrelated by critiquing the notion that the internet is a somehow separate place from the material world. They explore how social identity work flows through spaces and texts both online and offline. Davies (2007) further develops this line of argument, exploring how representations of our everyday selves and lives in online spaces such as Flickr.com (an online photo-sharing site) meld notions of public and private, online and offline, as narratives are woven through and around text and space. Based on this growing documentation of social identity work in online spaces, my interest lies in children's social identity play in these spaces, where identities are seen as multiple and connected to everyday local and global realities.

In addition, I am particularly keen to contribute to the field that aims to build connections between children's informal practices around digital technology and educational contexts. Research into children's digital text production often appears to be framed primarily by the educational domain. For example, in the United States, the Digital Youth Research Project proposal (Lyman, Ito, Carter, Thorne, 2006),[9] funded by the MacArthur Foundation since 2006, aims to explore the implications of children's innovative practices using digital media for schools and higher education. With this emphasis, research into children's digital text production is only ever viewed from within an existing educational framework. In this way, research that aims to fill gaps and meld practice from two fields becomes disproportionately influenced by the institution of the school and all of its complementary economic and political directions.

In England, children's text production continues to be a political concern, despite the efforts of the recent QCA E21 consultation (QCA, 2005) where a reinvented 'English' for the 21st century is imagined and described. Here, a significant body of researchers are calling for educators to reconceptualise notions of 21st century text (Bearne, 2005), and children's text producing behaviour (Carrington & Marsh, 2005). However, policy makers are only now beginning to respond to the radical sea change in children's text producing behaviours that have been afforded by mass access to the internet: an assertion that can be supported by scrutiny of the renewed Primary Framework for Literacy and Mathematics (DfES/PNS, 2006). However, although the screen is now included as a medium for reading and writing activities, schools have been reported as being reluctant to engage with children's informal online text producing habits.[10] Only if educators, researchers and policy makers work together to ensure that children's informal text production is valued and explored will new definitions of texts and their possibilities be realised.

Definitions of text have widened in the last 20 years in response to two separate but interconnected force-fields. First, notions of text have been expanded by the possibilities and affordances of new digital technologies. Screen-based, image-dominated texts are prolific in the new communication landscape. Adults and children navigate the technology required to create and receive screen-based texts as part of everyday life. Second, the increasing recognition of literacy as a social rather than merely skills-based practice has melded notions of literacy, text, discourse and communication (New London Group, 1996). The combination of these factors means that for text creators, educators and researchers, a 'text' is no longer just the 'lettered representation' conjured by a print-based, skills-driven literacy curriculum (Kress, 2003). Instead, our understandings of texts have grown to include combinations of words and images, sounds, gesture and movement. Texts are viewed as portals for the negotiation of social positioning and relics of our identities. Marsh (2005) describes how children of the digital age are shaped by and shaping new possibilities for communication and text production. She draws together a body of research that takes a new literacy studies perspective to frame how children's literacy practices are viewed (Marsh, 2005). A new literacy studies perspective contends that the literacy practices of individuals are 'inherently political and linked to issues of power, identity, inclusion and exclusion' (Carrington & Marsh, 2005). In this paradigm, social, cultural and personal components interplay to position individuals in relation to others (Holland & Leander, 2004). In this way, *identity, communicative practice* and *context* can be viewed as an inextricably linked triad which gives rise to the production of texts as the triad interplays. The texts produced are viewed as motivated signs (Kress, 1997) and artefacts of identity (Pahl & Rowsell, 2005, p. 108); spaces where social identities (Gee, 1996, p. 91) can be rehearsed, improvised and performed.

In an era described variously as late modern (Giddens, 1990), postmodern (Lyotard, 1984; Harvey, 1990) and liquid (Bauman, 2004), issues of identity are crucial, with some clearly arguing that humans construct and negotiate their identities discursively, in response to their histories and the sociocultural contexts within which they relate to others (Bauman, 2004; Gee, 1996; Marsh, 2005). Humans can be viewed in these terms as 'social actors' (Castells and Ince, 2003, p. 66) who invoke social identities as they communicate and draw upon their personal and cultural resources and histories (Holland, Lachicotte, Skinner, & Cain, 1998). Through this process, language is used and texts are produced.

Gee (1996) argues that language-in-use can be viewed as a scaffold for the performance of social activity and also as a scaffold for human affiliation within cultures, social groups and institutions. It is within this context that the notion of social languages and social identities is born. Gee describes how an individual enacts different social identities through their language-in-use on different occasions (Gee, 1999). He argues that language is not 'one monolithic thing', but rather is composed of many sublanguages—'social languages' that we use to communicate 'who' we are and 'what' we are doing in relation to the context we are in. As our social language in use varies from one context to another and within the same context, so we can recognise that humans are constituted of multiple 'whos' (Gee, 1996, p. 66). Our use of different social languages therefore makes visible our various social identities and can serve to either distance us from other participants, or to achieve solidarity and affiliation. The creation of a text, whatever the mode or modes involved, can be regarded as an artefact of this social identity work, whatever its materiality.

Holland et al. (1998) also present an explanation of social identity work that invokes a sense of negotiation. They describe how, in forming identities, humans produce understandings of themselves, about themselves and for themselves in relation to others. Significantly, this is seen as having the potential to guide subsequent behaviour.

> Identities are hard-won standpoints that, however dependent upon social support and however vulnerable to change, make at least a modicum of self-direction possible. They are possibilities for mediating agency. (Holland et al. 1998, p. 4)

Framing these theorists and researchers' work, lays Pierre Bourdieu's endeavour to find a theory for the social world (Jenkins, 2002). Bourdieu believed that the social use of languages reproduces systems of social difference. In *Language and Symbolic Power,* he describes how 'these styles, systems of difference which are both classified and classifying, ranked and ranking, mark those who appropriate them' (Bourdieu, 1991, p. 54). Bourdieu's observations predate the digital online contexts that Clare's negotiations of social identity and positioning now occupy. In addition, his discussion centres around speech and the subtle differences that pronunciation, diction and grammar realise for the sociologist who is observing the system of social differences. However, as Gee (2003) shows in his discussion of identity recruitment through video gaming, the positioning of ourselves in relation to each other through language use can also be applied to online contexts where a broader range of modalities and representative opportunities are available. Therefore, Bourdieu's observations about the social use of language to mark and rank ourselves amongst others, present an interesting platform from which to consider Clare's textual representations of herself in online profile pages and the social identity play that she is engaging in.

Both the pages in a traditional book-like photograph album and Clare's profile pages on a social network site can be viewed as repositories for social identity play. Davies (2006, 2007), in exploring the affordances of digital photo-sharing, explains that predigital photographs carry an iconic and artefactal status: they can be viewed and held; they inspire narratives that unfold around the image. Similarly, Pahl and Rowsell (2005), drawing upon Gee's large 'D' definition of Discourses, argue that texts must be seen as artefacts that carry traces of history, people and place. The construction of a traditional, book-like photograph album and the creation of a profile page on a social network site can be viewed as the creation of a textual artefact that requires the text-maker to choose representations of lives and selves for presentation to a range of audiences. These carefully selected representations provide opportunities for social identities to be realised and performed in different sorts of time and space.

To argue that a photograph album and profile page can be viewed as textual artefacts of social identity invokes a perspective of *text* taken from social semiotic theory (Kress, 2003). From this perspective, texts can be viewed as the 'stuff of our communication' fixed in a specific mode (Kress, 2003 and created within fields of power (Kress, 2003). This view in turn draws upon Bourdieu's vision of a social theory, where he argues that social identities are negotiated through practice across fields which depend upon and lead to social positioning (Bourdieu, 1991). Kress neatly defines text within this theoretical orientation: 'Text is the result of social interaction, of work: it is work with representational resources which realise social matters' (Kress, 2003). Both photograph albums and profile pages on a social network site can be considered from this perspective. They are artefacts that exist to position the creator in relation to the reader/audience. They are artefacts that have the potential to convey meaning. These artefacts are created as representations of the world as the author wishes it to be viewed and in ways that will produce the meaning that they wishes to communicate. They are artefacts that realise social matters (Kress, 2003) and as such are realisations of agency within a social world.

However, the opportunities for realising agency within online profile pages appear to differ from Atwood's traditional photograph album. A traditional photograph album can be viewed as a relatively permanent and chronologically shaped artefact. Some of the albums that occupy my mother's bookcase have not changed for over 50 years. Conversely, the profile pages on social network sites are highly ephemeral. Clare can edit and change the way in which she is represented at any time and for any reason. The profile page of a social network site is a fluid artefact of identity: one that morphs over time. Instead of presenting a series of contradictory images as a photograph album containing images taken over time might, the evolving profile page acts as a mask that the creator can choose to create and adopt. Subsequent masks replace the one before, as the creator edits and evolves the representations of herself in this space.

Bourdieu describes humans as agents who negotiate social relationships and positions within a social space, where the social world is constructed individually and collectively, based on the distribution of economic and cultural capital (Bourdieu, 1998). Central to this notion is Bourdieu's concept of 'habitus': the system of 'structured, structuring dispositions' (Bourdieu,

1990, p. 52) that circumscribes human behaviour and that 'function as structuring structures' (Bordieu, 1990, p. 53). For Bourdieu, the negotiation of social identities is a structural and agentive process, driven by the interrelationship of the habitus and the social field, in which the 'game' is being played. The field is defined by objective power relations and is the social arena in which negotiations and struggles occur (Bourdieu, 1991, p. 229). Within this, Bourdieu acknowledges that individuals may make calculated moves, but they are always defined and framed by the habitus which has been ingrained over time. Bauman (2004, p. 48) supports this view. For him, identity creation is a *means-orientated* rather than *goal-orientated* practice, where individuals draw upon the resources that they can make available in any given time and space.

If we subscribe to Bourdieu's notion of habitus and Bauman's means-orientated practice, we must accept that Clare's creation of online profile pages will reflect the structuring structures of her worlds as she creates and selects words, images and music to upload and represent herself to others. However, in a liquid modern society (Bauman, 2005), the structures and resources are evolving more rapidly than they can become habitual. Therefore, whether Clare is fixing her identity in time and space, or conjuring fleeting and changing representations of who she is as she plays with her desired social identity through online profile pages, is a question that bears exploration due to the possibilities offered in relation to the affordances, materiality and ownership of the texts produced.

My doctoral research is located within a new literacies studies perspective and involves interviewing preteenage children in order to consider how they negotiate their social identities in digital contexts. Clare engages avidly in many forms of digital text production and communication. In particular, Clare creates profile pages, originally on Bebo and now on MySpace to showcase who she is to the other people within her online social network. Clare's profile pages are usually composed of digital photographs, her words, video and music. In addition, these texts are contributed to by other people within her network who read and post comments. Clare regularly updates her profile pages, changing the images that are displayed and the text that is written. Two screen shots of Clare's MySpace profile pages that were taken six months apart illustrate how Clare's profile pages alter over time (Figures 5.1 and 5.3). These examples do not fully conjure her profile pages as the video clips that can be accessed, and the music that plays when the pages are opened are not able to be included here. In addition, the interactive features of the page where members have commented on her photographs cannot be displayed in full due to the extent of them. However, even from this inadequate presentation of data, it is clear that Clare uses visual images and words very carefully to construct her of-the-moment online identity. Below each screen shot, a section from her profile page, the 'About Me' blurb has been extracted and enlarged (Figures 5.2 and 5.4).

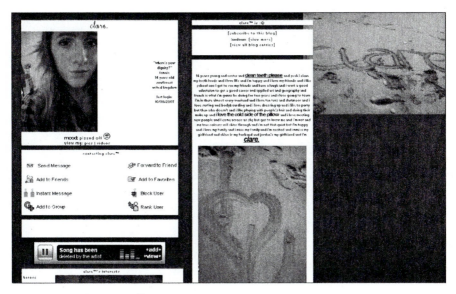

Figure 5.1   Clare's profile page, August 2007.

14 years young and exeter and **clean teeth please** and yeah I clean my teeth loads and I love life and I'm happy and I love my friends and I like school because I get to see my friends and have a laugh and I want a good education to get a good career and applied art and geography and French is what I'm gunna be doing for two years and I love going to town I'm in there almost every weekend and I love ten tors and dartmoor and i love surfing and body boarding and I love dressing up and i like to party but then who doesn't and I like playing with people's hair and doing their make up and **I love the cold side of the pillow** and I love meeting new people and I come across as shy but get to know me and I'm not and my true colours will shine through and I'm not that quiet but I'm happy and i love my family and i miss my family and I'm excited and immi is my girlfriend and Chloe is my husband and Jordan's my girlfriend and I'm
Clare.

Figure 5.2   Clare's 'About Me' blurb, August 2007.

Figure 5.3   Clare's profile page, February 2007.

These examples show Clare's play with her social identities over time. Her name in August is simply 'Clare', written in a cursive script; her mood: 'pissed off'; her tagline 'where's your dignity?' taken from Hilary Duff's latest album seen advertised on MySpace. In her blurb, she lists her current loves which include school, friends, family and the cold side of the pillow. Conversations with others are implied by her use of the seemingly random interjection 'clean teeth please' which the uninitiated reader can only presume refers to an in-joke or event from another time or place. Her profile photograph is arresting: she is staring directly at the camera, fully made up and open-eyed. This photo was taken in her older teenage cousin's bedroom, following a trip to a beauty counter for a makeover. The other images are moody sandscapes, carefully selected and presented in neutral tones. The overall effect is a sophisticated representation which reflects Clare's interests and friendships. That this page is used to affiliate her with older sophisticated friends and relatives is confirmed by reading the comments posted by others:

In February, Clare's profile name is 'Fart in a bottle' (see Figure 5.5). Her tag line: 'ask no questions, I'll tell no lies' is a well-known line used by her and her friends, taken from their own popular culture. This time her blurb contains two sections: an 'I love' list and a 'what my friends say about me' list, over which Clare has full editorial control. To compile this list, Clare requested comments from her friends during MSN Instant Messaging sessions. She then selected and added the comments that she liked to her profile page. Her photograph is more coy than in August: Clare's eyes are downcast and her eye make-up and profile are emphasised. Clare explains that lots of her friends take photographs like this for their MySpace profile pages.

In both pages, the attention given to artistic presentation and the desired impact on the audience is significant. These pages are constructed carefully

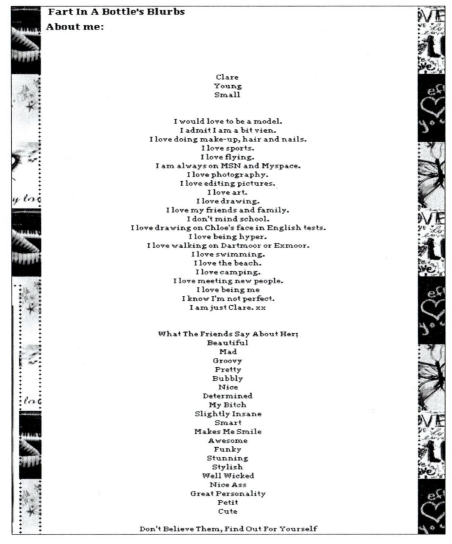

Fart In A Bottle's Blurbs

About me:

Clare
Young
Small

I would love to be a model.
I admit I am a bit vien.
I love doing make-up, hair and nails.
I love sports.
I love flying.
I am always on MSN and Myspace.
I love photography.
I love editing pictures.
I love art.
I love drawing.
I love my friends and family.
I don't mind school.
I love drawing on Chloe's face in English tests.
I love being hyper.
I love walking on Dartmoor or Exmoor.
I love swimming.
I love the beach.
I love camping.
I love meeting new people.
I love being me
I know I'm not perfect.
I am just Clare. xx

What The Friends Say About Her;
Beautiful
Mad
Groovy
Pretty
Bubbly
Nice
Determined
My Bitch
Slightly Insane
Smart
Makes Me Smile
Awesome
Funky
Stunning
Stylish
Well Wicked
Nice Ass
Great Personality
Petit
Cute

Don't Believe Them, Find Out For Yourself

Figure 5.4   Clare's 'About Me' blurb, February 2007.

using available websites and knowledge-sharing amongst peers. The result is impressive, particularly as the skills to create MySpace and Bebo profile pages are learnt informally and seemingly effortlessly, with and from friends and the social network sites themselves. Clare's use of these sites appears to be driven by her enjoyment of playing with her representation of self and the agency that she derives from this. In this sense, the creation of her profile page seems to be a goal-orientated performance of identity in a way that resembles the selection and inclusion of a photograph in a

Figure 5.5   Comments from Clare's profile picture, August 2007.

traditional album. However, three features of Clare's profile pages impact on this process to make it an improvised rather than polished performance: the affordances of the online profile pages, their materiality and the qualities of ownership and authorship.

The production of any textual artefact, such as a book-like family photograph album or an online profile page, occurs within a landscape that is circumscribed structurally by the available resources, communicative practices, and their affordances. For Atwood's or my mother's baby boomer generation, the traditional family photograph album was typically a book-like text, where photographs were discriminately taken, developed, selected and glued or inserted into an album. Sometimes handwritten headings and annotations supported the images.

Here, in Figure 5.6, my mother's identity is performed and fixed in time and space. While the audience in relation to the photograph may change, my mother is archived as a dancing child, in a flower-filled back garden of a terraced South-London house, on Coronation Day 1953. Narratives which help to construct my mother and our family's history, as they are viewed and reviewed, exist around this image.

In the first decade of the 21st century, the production of texts that realise social matters is a very different affair (Kress, 2003). Issues of

Figure 5.6   A page from my mother's photograph album, 1953.

audience, materiality and ownership mean that Clare is unlikely to retain such a concrete textual archive of social identity to be reviewed over time. The evolution and mass uptake of digital technologies and internet access have revolutionised text producing and consuming behaviour in creative and economic terms. In relation to 'the new media age', Kress portrays these contexts as 'semiotic worlds' (Kress, 2003, p. 4); spaces where the medium of screen and mode of image are dominant, and where text production and consumption is increasingly driven by the affordances of new digital media. Within Kress's semiotic worlds, representations of social identity are shifting from a micro level to the macro. Textual artefacts, such as family photographs or home videos, previously available to small selected micro audiences, can now be published and distributed widely and sometimes indiscriminately using social network sites. Clare currently has 96 'friends' listed on her profile page who can access her site. In addition, these friends have left 606 comments in relation to her blurb and profile. At a macro level, Kress argues that these shifts in affordances give the creator of digital textual artefacts the potential for unprecedented authority and power.

It is apparent from talking to Clare about her 'friends' list that her profile pages are reaching both local and global audiences. While many of her 'friends' are known personally to her through school, a large number of listed contacts are friends of friends or relatives; children who are older or younger; and who live across a widespread and local geographical location. Clare's profile pages are therefore both enlarging and ring-fencing her social world. In this way, the macro and micro level of audience can be seen to exist together. Consequently the field (Bourdieu, 1990) in which Clare plays, as she negotiates who she is and how she is positioned, is equally extended and contained. The affordances of MySpace provide Clare with the liberty to improvise aspects of her social identities more playfully than if the field was smaller. However, Clare is not free to contrive an entirely altered image as her field also includes anchors to her local and day-to-day self. Carrington explores the potential for transformation of self in textual landscapes (Carrington, 2005b, p. 20). She describes how children inhabit, create and are positioned within 'textual landscapes' through which they continually interact consciously and subconsciously with a range of texts to express their identity. Carrington re-emphasises the individual within these textual landscapes. She notes how the landscapes circumscribe the individual's interactions with texts and are equally circumscribed by them. She reminds us that 'the agentive child engages purposefully in these textual landscapes as a means of expressing her/his identity and place in the social world' (Carrington, 2005b, p. 21). Clare's presentation of social identities can be seen at once to be structured and structuring within her online profile pages. As such, they cannot be viewed as just polished performances, but also as improvisations that are replayed and reworked in response to local and global, micro and macro influences.

Equally, the materiality of these online texts is very different to the materiality of texts produced using more traditional methods. Traditional methods—either the construction of school-based texts using pencils and paper, or the creation of a photograph album where photographs are developed and arranged in order—render communication and representation in a concrete form. In these processes, ideas and personas are fixed in time and space. Through this fixing, the intentions and capabilities of the author are captured too, leaving a trace of who we are and what we think for others and ourselves to return to and reflect upon—albeit from new angles and perspectives. Even the digital texts produced in school using commercial programmes are subject to this fixing, as students save their texts for marking and appraisal by their tutors or peers.

Online profile pages are different in this respect. They are transient and mask-like. Instead of fixing representations of who we are as photos in an album or a saved document might, newly edited profile pages on MySpace simply replace earlier creations. The images, words, music and movement selected to represent the author can be revised at any time. Previous versions are overlayed in this process and vanish from display. In this way, the texts created as profile pages have the potential to be truly ephemeral. The author can choose to change how they appear in these online spaces as easily as they change their clothes. That this may sometimes confirm and sometimes contradict the author's 'normal' everyday appearance is simply part of the fun to be had with the audience.

However, although profile pages afford their authors the opportunity to trade appearances, they also have the potential to exist as concrete artefacts with an enduring material presence. Profile pages can be saved and captured by the author or audience as screen shots. These texts can be reviewed as a photo in a traditional album might. Narratives can be built around such an image that is captured and saved in much the same way as they can for a photograph. This can have consequences for the author. A recent example occurred where the proctors at Oxford University used captured entries found on the social network site Facebook as evidence of antisocial behaviour in a disciplinary hearing relating to an end of term 'trashing' incident.[11]

A third feature that impacts on the notion of online profile pages as sites for the improvisation alongside the polished performance of social identity is the notion of ownership. By ownership, I am referring to the issue of authorship in spaces where other people can comment and contribute. For example, on Clare's profile page, readers can post comments in response to her postings at any time. This means that the profile page that is originally created by one person may, over time, accrue more voices and acquire a polyvocality. The comments contribute to the text as a whole and add texture to the representation of the individuals involved. In this way, the text ceases to be created by one author, and instead becomes a reflection of a networked writer in negotiation with others. Clare allows comments on her profile pages, and in doing so, relinquishes full editorial control of a textual artefact that reflects her social identity. This represents a shift from the kinds of authoring and co-authoring that are made possible by traditional paper-based writing activities, where ownership of the process is usually retained by one main author, or controlled by established boundaries, such as the end of a lesson, or the end of a play episode. This feature contributes to the notion that Clare's texts are not entirely polished performances of social identity, but are artefacts of social identity in the improvisation, making and remaking

Throughout this discussion I have tried to explore the idea that as a producer of textual artefacts that contribute to her positioning within her social world, Clare both performs and improvises her social identities. Instead of creating and performing an archive of her various desired social identities within a photograph album as my mother, or Atwood's character described in the opening section of this chapter have, Clare engages in dynamic online text production and communication that is continually evolving, leaving a trajectory of Clare that is always topped by the most recent and desired impression. This continual reworking of social identity through text can be seen as a by-product of Bauman's liquid life: where self-scrutiny, self-critique and self-censure are constantly applied as we pursue unattainable happiness in a liquid modern society (Bauman, 2005, pp. 11–12).

An aim of this chapter was to stand away from existing educational frameworks and explore what it is that children do when they produce texts in the new communication landscape. The notion that children do not only perform but also improvise social identities throughout the creation of online profile pages has been explored. This redirects our attention from the idea of text producing as a process and product merely to be assessed in educational settings towards a wider interpretation of text creation as a social endeavour that involves a spectrum of performance—from polished to improvised.

This chapter only attempts to reflect the idiosyncratic actions of one child's online text making. However, by considering her actions within this alternate frame, it is hoped that wider issues around children's text production will be raised within the 21st century communication landscape. In describing this landscape, Kress (2003) and Carrington (2005a) conjure scenes of children and adults engaging with screen-based, image-driven digital textual communication as part of their everyday lives in order to achieve social positioning. Here, positioning through digital textual representation combines the microprocesses (involving the small world agency of the text producer within their local environment) and the macroprocesses (involving the affordances of new technologies and media) to achieve new forms of communication and representation that differ significantly from the baby boomer photograph albums described by Atwood. My mother embraces the increased opportunities afforded by new technologies as she takes digital images which she deletes or saves, edits, emails, and occasionally prints out. She views digital photographs on screen, and uses the internet to exchange photos with friends and family. However, her digital activities remain framed by her small-world microexperiences of taking predigital photographs and constructing the book-like family photograph album; creating a crafted composition that is fixed temporally and spatially, and that presents solid versions of self and family which are shown to a limited selected audience and located within the predigital textual landscape of the bookcase in the family home. As such, my mother represents the generation of digital immigrants (Prensky, 2001) who view text production as a performance of literacy and academic achievement. What I have raised through this discussion is the notion that the textual artefacts of the digital age that children are creating informally and prolifically differ significantly from the predigital textual products that could be seen as portals for social identity work. Clare's profile pages are simultaneously sites for polished performance and improvisation; they are both one-voiced and many-voiced; they are both transient and permanent; they are both meaningful and meaningless. Consequently, those responsible for supporting children as they learn to create texts in this communication landscape need to recognise the widening spectrum of textual possibilities and text-producing behaviours where texts function to position children in relation to each other in complex and multiple ways.

NOTES

1. MSN Messenger is a service that is owned by the Microsoft Corporation. It was renamed MSN Live Messenger in February 2006.
2. The UKCGO ESRC and e-society research project was conducted between April 2003 and April 2005; 1511 children between 9–19 years were surveyed. For details of the sample and methodology, see the final report at http://www.children-go-online.net/. According to this project, 75% of 9–19-year-olds have accessed the Internet from a computer at home, and 38% of these children have a mobile phone.
3. The Office for National Statistics online reports that: 'In 2007, nearly 15 million households in Great Britain (61 per cent) had Internet access. This is an increase of just over 1 million households (7 per cent) over the last year and nearly 4 million households (36 per cent) since 2002. . . . Eighty four per cent of UK households with Internet access had a broadband connection in 2007, up from 69 per cent in 2006.' Sources: National Statistics Omnibus Survey; Northern Ireland Omnibus Survey; Survey of Internet Service Provider. Published on August 28, 2007 at 9:30 a.m.
4. For access to Bebo, please visit http://bebo.com/
5. For access to MySpace, please visit http://www.myspace.com/
6. See http://www.danah.org/papers/WhyYouthHeart.pdf. for boyd's overview of social network sites and profile pages.
7. For access to Alexa.com, please visit http://www.alexa.com/site/ds/top_sites?cc=GB&ts_mode=country&lang=none.
8. See BBC Video and Audio reports search page http://search.bbc.co.uk/cgi-bin/search/results.pl?tab=av&q=myspace&recipe=all&scope=all&edition= and BBC Video Rise of Social Networks http://news.bbc.co.uk/player/nol/newsid_5390000/newsid_5396000/5396020.stm?bw=bb&mp=wm&nol_storyid=5396020&news=1 (Retrieved July 28, 2007) for media reports detailing the extensive uptake of social network sites such as MySpace and Bebo.
9. For details of the ethnographic research studies that are contributing to the Digital Youth Project see http://digitalyouth.ischool.berkeley.edu/node/1.
10. For examples of media reports about schools' concerns with popular social networking sites see http://search.bbc.co.uk/cgi-bin/search/results.pl?tab=av&q=bebo&recipe=all&scope=all&edition= (Retrieved October 7, 2007).
11. For details of this incident see Guardian Newspaper Online http://www.guardian.co.uk/g2/story/0,,2129494,00.html (Retrieved July 15, 2007).

REFERENCES

Alexa. (2007). [Web information company]. Top sites United Kingdom. Retrieved July 28, 2007, from <http://www.alexa.com/site/ds/top_sites?cc+GB&ts_mode=country&lang=none>

Atwood, M. (2006). *The tent.* London: Bloomsbury Publishing.

Barton, D., & Hamilton, M. (2000). Literacy practices. In. D. Barton, M. Hamilton, & R. Ivanic (Eds.), *Situated Literacies,* (pp. 7–15). London: Routledge.

Bauman, Z. (2005). *Liquid life.* Cambridge, UK: Polity Press.

Bauman, Z. (2004). *Identity: Conversations with Benedetto Vecchi.* Cambridge: Polity Press.

Bearne, E. (2005). Interview with Gunther Kress. *Discourse: Studies in the cultural politics of education,* 26(3), 287–299.

Bourdieu, P. (1990). *The logic of practice*. Cambridge, UK: Polity Press.

Bourdieu, P. (1991). *Language and symbolic power*. Cambridge, UK: Polity Press.

Bourdieu, P. (1998). *Practical reason*. Cambridge, UK: Polity Press.

boyd, d. (2007). Why youth (heart) social network sites: The role of networked publics in teenage social life. *MacArthur Foundation Series on Digital Learning, Identity Volume*. (D. Buckingham, Ed.). Retrieved July 23, 2007, from <http://www.danah.org/papers/WhyYouthHeart.pdf >

Carrington, V. (2005a). New textual landscapes. In J. Marsh (Ed.), *Popular culture, new media and digital literacy in early childhood* (pp. 13–27). London: RoutledgeFalmer.

Carrington, V. (2005b). Txting: the end of civilisation (again). *Cambridge Journal of Education, 35*(2), 161–175.

Carrington, V., & Marsh, J. (2005). Digital childhood and youth: New texts, new literacies. *Discourse, 26*(3), 279–285.

Castells, M., & Ince, M. (2003). *Conversations with Manuel Castells*. Cambridge, UK: Polity Press.

Davies, J. (2006). Affinities and beyond! Developing ways of seeing in online spaces. *E-Learning, 3*(2), 217–234.

Davies, J. (2007). Display; identity and the everyday: Self-presentation through online image sharing. *Discourse: Studies in the Cultural politics of Education, 28* (4) 549–564.

Department for Education and Skills/ Primary National Strategy. (2006). *The primary framework for literacy and mathematics*. London: DfES/PNS.

Dowdall, C. (2006a). Ben and his 'army scenes': A consideration of one child's 'out-of-school' text production. *English in Education, 40*(3), 39–54.

Dowdall, C. (2006b). Dissonance between the digital words of school and home. *Literacy, 45*(3), 153–163.

Gee, J. (2003). *What computer games have to teach us about learning and literacy*. London: Palgrave McMillan.

Gee, J. P. (1996). *Social linguistics and literacies (second edition)*. London: RoutledgeFalmer.

Gee, J. (1999). *An introduction to discourse analysis*. London: Routledge.

Giddens, A. (1990). *Modernity and self-identity*. Cambridge, UK: Polity Press.

Harvey, D. (1990). *The condition of postmodernity*. Oxford, UK: Blackwell.

Holland, D., & Leander, K. (2004). Ethnographic studies of positioning and subjectivity: An introduction. *Ethos, 32*(2), 127–139.

Holland, D., Lachicotte, W., Skinner, D., & Cain, C. (1998). *Identity and agency in cultural worlds*. Cambridge, MA: Harvard University Press.

Jenkins, R. (2002). *Pierre Bourdieu revised edition*. London: Routledge.

Kress, G. (1997). *Before writing: Rethinking the paths to literacy*. London: Routledge.

Kress, G. (2003). *Literacy in the new media age*. London, UK: Routledge.

Kress, G. in Bearne, E. (2005) Interview with Gunther Kress, *Discourse: studies in the cultural politics of education*, Volume 26 (3) 287–299.

Leander, K., & McKim, K. (2003). Tracing the everyday 'sitings' of adolescents on the Internet: A strategic adaptation of ethnography across online and offline spaces. *Education, Communication & Information, 3*(2), 211–240.

Lewis, C., Fabos, B. (2005). Instant messaging, literacies and social identities. *Reading Research Quarterly, 40*(4), 470–501.

Livingstone, S., & Bober, M. (2005). *UK children go online final report of key project findings*. London: Department of Media and Communications, London School of Economics.

Lyman, P., Ito, M., Carter, M., and Thorne B. (2006). *Digital youth research.* Retrieved October 8, 2007, from http: <http://digitalyouth.ischool.berkeley.edu/node/1>

Lyotard, J. (1984). *The postmodern condition: A report on knowledge.* (G. Bennington & B. Massumi, Trans.) Manchester, UK: Manchester University Press.

Marsh, J. (2005). Children of the digital age. In J. Marsh (Ed.), *Popular culture, new nedia and digital literacy in early childhood.* London, New York: RoutledgeFalmer.

National Statistics Office. (2007). *National Statistics Omnibus Survey; Northern Ireland Omnibus Survey; Survey of Internet Service Provider.* Retrieved February 11, 2007, from http://www.statistics.gov.uk/cci/nugget.asp?id=8

New London Group. (1996). A pedagogy of multiliteracies: Designing social futures. *Harvard Educational Review, 66*(1), 60–92.

Pahl, K., & Rowsell, J. (2005). *Literacy and education.* London: Paul Chapman Publishing.

Prensky, M. (2001). Digital natives, digital immigrants. *On The Horizon, 9*(5), 1–6.

Qualifications and Curriculum Agency. (2005). *English 21: Asking the questions to shape the future of English.* Retrieved November 18, 2005, from http: <http:www.qca.org.uk/11775.html>

Thomas, A. (2004). Digital literacies of the cybergirl. *E-Learning, 1*(3), 358–382.

6 Exciting Yet Safe
The Appeal of Thick Play and Big Worlds

Margaret Mackey

A number of years ago, I was asked to proofread a publicity brochure for the Faculty of Education at my university, along with my American colleague Jill McClay, a long-time resident of Canada. We had not read much of what I perceived as its very innocuous prose when she broke down in gales of laughter. 'Every time I think I finally understand Canadians, I see something like *this*', she gasped when she finally got her breath back. I looked carefully at the brochure but nothing struck me as unusual. She pointed. 'The city of Edmonton is exciting yet safe', promised the blurb.

'Any American knows that's an either–or proposition', said Jill. 'Only Canadians would think you could have both excitement and safety at the same time'. She laughed harder.

I do not propose to explore who is right in this national-existential divide, but I do know that there is a different group of people who know a great deal about the pleasures of 'exciting yet safe'. These people are readers, viewers, players, who commit themselves to a fictional world and return to it over and over again. They have the security of knowing that world and its limits, along with the excitement of novelty in the details as they encounter new elements of the story. The theme of enjoying a fictional world intensely and of returning to it in a variety of ways has dominated many of the 70 interviews I have conducted with readers over the past 15 years. Over that period of time, significant advances in domestic technologies have developed and reinforced an emphasis on return engagements in popular culture. DVDs build on the repetitive capacities of videocassettes and the added tracks provide new ways of engaging with a film. Websites, chatrooms, listservs and other digital forums provide a vast repertoire of background information and fan reworkings of loved fictions in all media.

In this chapter, I explore some of the implications of the kinds of intense relationship that develop when people return over and over again to the same fictional universe. In this discussion, I draw on numerous conversations with readers, viewers and players, and making use of two ideas that I developed from some of these conversations: the idea of thick play and the concept of the big world (Mackey, 2007).

I refer to intensive and extensive play as 'thick' in the sense that Clifford Geertz (1975) talks about 'thick' description as being full, dense, rich and complex. An ethnographer aiming at thick description must deal with a messy world that contains 'a multiplicity of complex conceptual structures, many of them superimposed upon or knotted into one another, which are at once strange, irregular, and inexplicit' (1975, p. 10). Thick play offers ways of lingering in a particular fictional world, savouring, repeating, extending and embellishing the imaginative contact with that world, often in complex, irregular and inexplicit ways that may indeed be 'superimposed upon or knotted into one another'.

Thick play describes a form of activity; 'big worlds' are often where that activity takes place. In my terms, a big world is a fiction that extends beyond the limits of one text. Big worlds can be found in many formats. A set of series books or sequels involves a big world. So do adaptations that tell the same, or a very similar story, in different ways. Many computer games create big worlds, and the multiplayer online digital games are not simply big but ever-expanding, as players create new content every day; the support materials that players also create for themselves and each other move the fictional encounter beyond the limits of the single title.

In fact, contemporary popular culture fosters big worlds in a variety of ways. The extensive support structure that accompanies the release of new movies or television shows, and that we now all take for granted, is a kind of world-enlarger. The apparatus of add-ons provides scaffolds to the original story in the form of websites (some of which expand on the fictional universe), magazine specials, TV interviews and behind-the-scenes shows, novelizations and picture books, toys, collectibles and consumables, associated digital games, and so forth. The cross-promotion of movies with other products is so ubiquitous that it sometimes feels as if the small world fiction is the exception.

The production of a big world does not automatically create the engrossed interest that leads to thick play. When a big-world story does catch the imagination of a reader or viewer or player, however, it does offer a variety of routes into absorption. Thick play often involves what Mihalyi Csikszentmihalyi calls 'flow' experiences, where actions seem effortless and automatized and attention focuses on content rather than on process (1997, p. 29). 'When goals are clear, feedback relevant, and challenges and skills are in balance,' says Csikszentmihalyi, describing flow, 'attention becomes ordered and fully invested' (1997, p. 31). Not all thick play is in the flow zone, however; much of it involves what we might call preflow, the kind of extensive practice in the teeth of frustration that thick players must invest in order to get to that level of automaticity where it all seems effortless and attention can focus purely on content rather than the means of access.

Thick play and big worlds come in many guises. In the following sections, I present some examples from readers, viewers, gamers and explorers whom I have interviewed over the years. As well as citing the comments of children, I include many remarks by adults in their 20s. In a time of rapid technological change, the definition of 'exciting yet safe' is distinctly generational. I am a reader who learned to decode print before ever seeing a television set. As a result, there are many forms of digital text that I will never perceive as 'safe'; and what seems exciting to a digitally informed reader, often seems quite daunting, even threatening, to me. Knowing the gaps in my own history, I find that my understanding is enhanced if I refer to a full spectrum of 'new literates', from childhood to early adulthood, in order to gain some real understanding of how thick play and big worlds function in contemporary multimodal literacy.

BUILDING FICTIONAL COMMITMENT

Revisiting the Same Text

There are many ways of returning to a fictional world. Not all thick play necessarily entails big worlds. The simplest route of return is simply to reread (or rewatch, or replay). Thirteen-year-old Candace kept a list of her very favourite books, where she recorded her rereadings; at the time of our conversation, she was up to 23 readings of *The Secret Garden* (the list also contained *The Sky is Falling* by Canadian author Kit Pearson, and the five titles of Lloyd Alexander's *Chronicles of Prydain*; [Mackey, 1995, p. 168]).

Candace allowed for some interval between readings; 21-year-old Ben was even more committed to his favourites. For example, he watched the DVD of *Batman Forever* 'every day for about two months', committing large swathes of dialogue to memory (Mackey, 2007). At the time of our meeting, he had just acquired the newly released DVD of *Spider-Man* and was planning an equally intense immersion. Ben did not plunge into every text he encountered in this extreme way but he clearly had a proclivity for thick-play loyalty and was prepared to commit himself wholeheartedly to materials that caught his fancy.

Barbara Klinger writes about the pleasures of re-viewing movies on DVD, and points out how intense exposure to a text weaves it into personal history.

> Repetition amplifies any domestic medium's ability to become part of viewers' daily lives, even part of their autobiographies, resulting in an intense process of personalization. Like other objects, films experienced repeatedly in the home can attain an intimate, quasi-familial status that affects their meaning and influences individuals' perceptions of themselves and the world. (2006, p. 139)

Digital affordances allow for forms of re-engagement with variations; online discussion groups of television programs often indulge in almost line-by-line recapitulation of a viewing experience, along with commentary and critique. Full screenplays, posted online, sometimes provide even more detailed opportunities for revisiting a text. New technologies are also expanding the possibilities of return to a story. For example, the digital game *Halo 3*, nearing its release time as I write, will enable a player to make a personal recording of his or her game highlights *as played*, offering a sophisticated way to return to a known fiction. Here is Wikipedia's advance description of the powers of this feature:

> *Halo 3* will have a feature called *Saved Films*. This feature allows players to save a copy of the game data of a multiplayer match or campaign session to their Xbox 360's hard drive, so that they may watch it later on. . . . Players will be able to view the action from almost any angle and any player's perspective (including a free-roaming camera), as well as being able to slow down the speed and also play the recording in reverse. The Saved Films can even be edited in game to create a shorter clip of a particularly amazing or special moment. Players can also use the tool to take still pictures from films. (http://en.wikipedia.org/wiki/ Halo_3#Cooperative_play, retrieved August 26, 2007)

Just as video and later DVD transformed the evanescent nature of film-watching, so such features as the Saved Film function will render computer games revisitable in new and interesting ways. The power of being able to return is significant, and it is fascinating to see the digital game industry beginning to explore and exploit that power.

Series Reading

Candace did not explain the compelling attraction of *The Secret Garden*, but she was more articulate about an expired passion for *The Baby-sitters Club*, a relatively anodyne series of books for preteens by Ann M. Martin.

> I still read them because I hate them so much. I'm looking for one where something bad happens to one of them. [laughs] I haven't found it yet. . . . I'm looking for one that, that, where—I want one of them to die. Stacey has diabetes; I'm waiting for her to drop dead of diabetes. (Mackey, 1995, p. 169)

Clearly for Candace, the excitement of *The Baby-sitters Club* has worn off and the series is now too safe to be appealing. But in general, the series that successfully appeals to a reader or viewer is perhaps the classic incarnation of 'exciting yet safe'. Readers feel at home in the world, yet there are new situations, adventures, threats and surprises to be considered.

Candace may have outgrown *The Baby-sitters Club* but series reading is not merely a childhood experience. Ben, who clearly shared some of her predilections for repeat engagements, was a committed reader of Terry Goodkind's series, *The Sword of Truth*. When a new title appeared, he bought the hardback on the day of its release and abandoned all other entertainment until he finished it, reading on the bus, in bed, and surreptitiously while he was supposed to be paying attention to work or school.

Ben's experience of Goodkind's world thus includes an enforced waiting time for the next book to arrive, a consequence of serial consumption that inflects encounters with books, television programs and other forms of sequel in ways that have not received much consideration. Anticipation, speculation and prediction fill the gap of waiting time and expand the emotional reach of a series over an extended time, years in the case of *The Sword of Truth*—or, indeed, *Harry Potter* or *Halo 3*. *Harry Potter* fans, waiting for the final volume of Rowling's series to appear, devoted websites to making predictions about the final outcome of the story; some of these predictions were gathered together and published in book form (Schoen, 2006)—an extreme example of expectancy, perhaps. Anticipation of *Halo 3* also filled websites and featured in a cover article in *Wired* Magazine (Thompson, 2007). We do not know enough about the impact of such anticipation but we do know that reading a series over its real time of publication is different from being able to read it all in one go. Similarly, watching an entire television series on DVD is not the same experience as waiting for the weekly instalment on the box (Mackey, 2006). Both are big-world experiences but they are felt differently.

Different readers deal differently with series books. Some are happy to read promiscuously out of order, depending on availability. Others are not. Ten-year-old Kristen, who had just read *Nancy Drew #1* when she talked to me, is a believer in reading books in order: 'I just think if I skip a book or something, I'd probably miss something'. Kristen's reading life is made more complicated by this taste for the correct sequence, but she has developed a variety of acquisition strategies. Talking about *The Baby-sitters Club*, she said: 'Well, my grandma bought the first five for me and then my mum bought me some, but I get now, most of them from the library and if they're not here then I borrow them from a friend or something'. If she were looking for #32 and only #33 was in the library, she would not pick it up, but would pursue it through other channels. As with Ben's careful accumulation of the Goodkind titles, the rituals of acquisition appear to form a part of the thick play surrounding the books.

Marketers are not slow to leap on the anticipation bandwagon, but the psychological build-up to the release of a title in any medium that will expand a big world is not just a marketing phenomenon. Informed anticipation entails interpretive scrutiny of previously published components of that world (what clues are present in the first six books about Harry Potter and what significance should be ascribed to them?) and is a source of pleasure in its own right.

Adaptations

The re-encounter and the serial encounter are not the only vehicles for returning to a known fictional universe. Another form that enables repeat engagements is the adaptation.

Linda Hutcheon raises the question of why adaptations are such a strong element in our culture.

> If adaptations are . . . inferior and secondary creations, why then are they so omnipresent in our culture and, indeed, increasing steadily in numbers? Why, even according to 1992 statistics, are 85 percent of all Oscar-winning Best Pictures adaptations? Why do adaptations make up 95 percent of all the miniseries and 70 percent of all the TV movies of the week that win Emmy Awards? . . . [T]here must be something particularly appealing about adaptations *as adaptations*. (2006, p. 4)

Jill McClay, the baffled American expatriate, would recognize the elements in Hutcheon's answer to this question, which clearly includes the concept of 'exciting yet safe':

> Part of this pleasure, I want to argue, comes simply from repetition with variation, from the comfort of ritual combined with the piquancy of surprise. Recognition and remembrance are part of the pleasure (and risk) of experiencing an adaptation; so too is change. (2006, p. 4)

Peter Lunenfeld calls ours 'an aesthetic of 'unfinish'' (2000, p. 8) and talks about both commercial and cultural motivation for the current propensity to return over and over again to the same story. Today's young people have grown up in this world of the never-ending story, the fiction that is reworked again and again in a variety of media, analogue and digital, paper and screen. They become experts not only in the world of the story but also in the art of adaptation.

Every adaptation is an interpretation, and young people who move from one version to another develop at least a tacit sense of different ways of reading the same story. Phyllis Bixler, discussing Marsha Norman and Lucy Simon's stage musical version of *The Secret Garden*, talks about the significance of misreading. This musical draws on a chorus of ghosts to help tell the story of the motherless Mary and Colin. Bixler talks about the significance of the ghosts to one interpretation of the story, and also mentions their position as providing some practical solutions to some problems of adaptation:

> In addition to providing the non-linear, nonliteral richness that the team wanted . . . the ghosts solved a number of musical and theatrical problems. They provided a singing chorus. . . . The ghosts also solved

some dramatic problems, such as the cholera epidemic, which had earlier been abandoned as unstageable. . . . Perhaps above all, the ghosts allowed the musical to be faithful to what Norman has called a crucial 'promise' that Burnett's book offers . . . the promise that 'no matter where I am, no matter what happens to me, I will always be with you and you will be all right.' (1994, p. 107)

Even this brief quote gives some sense of this kind of big world as a sort of palimpsest, where one interpretation layers upon another, and where a different kind of ghost—the ghost of a familiar but not identical story—may haunt the play, especially perhaps for those who have read the original book with eyes blind to that 'crucial' promise that Norman has discerned in Burnett's writing.

The appeal of exploring different versions of a story is omnipresent in contemporary culture, and is a pleasure understood by quite young readers. Seven-year-old Barbara told me she likes to look at a movie and a book of the same story. 'Well, sometimes it's different, so I just want to see what's different about it. I find it interesting, finding out, like, about the same thing but it's made differently.' Two friends, Susannah and Alison, spent a part of their summer holiday in the year they were twelve systematically pursuing versions of *Little Women*: reading the book, watching a couple of movie versions and listening to an audio tape. Asked which they liked best, they paused for reflection before Susannah said, 'Well, the book is bigger'. That sense of investing 'thickly' in a story is clearly a source of pleasure, and contemporary domestic media enable many forms of thick encounter. Readers do vary in their enthusiasm for this kind of engagement, however; Kristin, age 10, said she was leery of seeing a movie of a book she really enjoyed: 'Normally the movies are always different from the book so I wouldn't want to see the movie'.

Domestic technology makes exposure to adaptations commonplace and even those, like Kristin, who are leery of a new version have broad general understandings of how adaptations work. For many contemporary readers, they provide intriguing opportunities to rethink a familiar story.

SELECTIVE PASSIONS

At the same time as they relish some forms of return to a loved fictional zone, and refine their capacity to criticize how different incarnations have been created, readers, viewers and players may choose to be very selective in how they attend to different instantiations of and sequels to a loved story. A former student of mine offers a detailed example of such discriminatory passion at work.

Warren Maynes is a long-time aficionado of *Highlander*, a fiction that appeared in a variety of formats: films, novels, television series, card games, and websites. Over 20 years, his emotional connection to the story was a significant part of his life but he reacted very differently to different sequels and versions of the story. The original movie (*Highlander*, 1986) caught his imagination as an adolescent; he was captivated by the idea of the existential hero using his sword to defeat the foe and gain all the wisdom and experience of that vanquished enemy through the 'quickening' that followed victory. But the first film sequel (*Highlander 2: The Quickening*, 1991) appalled him; he found it 'a traumatic experience' that gave him the 'dry heaves' (Maynes & Mackey, 2006, n.p.), as the movie toyed carelessly with the rules of survival that had so engaged him on the first encounter with the story. At first, scarred by this experience, he vowed to watch the subsequent television series only for the sword fights, but gradually he became absorbed in the stories once again. The culmination of his commitment to this saga, however, came with the production of a Collectible Card Game, a form of the story to which he dedicated years of Saturday mornings (Maynes & Mackey, 2006, n.p.). With the card game, he found he could apply the rules that so pleased him to the entire fictional universe of this story and regain some control over his relationship with different characters and episodes. He used the web to develop further cards when the manufacturer stopped making them.

Warren is not alone in his selective obsession with different versions of a particular story, although the fact that his commitment lasted for two decades perhaps puts him at an extreme end of a well-populated spectrum. Nor is this kind of passion confined to the popular culture end of the scale. Devotees of Jane Austen frequently favour one television adaptation over another, often quite vociferously. The power of discrimination among various renditions of Shakespeare carries considerable cultural prestige.

Drew, age 28, offers another example of selective commitment. Over a period of years, he engaged with versions and descendants of *The Lord of the Rings* in a huge variety of ways, watching the films over and over again, listening intensively to the soundtracks, viewing all the extra tracks on the DVDs and reading the liner notes on the CDs, reading Tolkien-derived *Forgotten Realms* fantasy novels and playing a fantasy game with friends based on the *Forgotten Realms* world. Yet when he attempted to read the books themselves, he found he could get interested only in the appendices that describe how Tolkien created his world; the actual story was 'too wordy' and he could not find a way to immerse himself in it.

THE APPEAL OF BEING THE EXPERT

There is an appeal in becoming an expert on a particular fiction. Young people have for generations plunged themselves into particular fictions and explored them as extensively and obsessively as the range of available texts made possible. Today, however, potential sources of information about some popular fictions have multiplied exponentially: It is easy to find background information on most movies, for example, on the extra tracks of the DVD version, in entertainment television programs, on the Internet, in specialist magazines and in books about the making of the movie.

Print fiction may also exist within the context of a strong support system of supplementary materials. Free-standing novels do not necessarily attract the same elaborate scaffolding, but many series books are well supported by outside materials, most often online. For example, it is always an education to check out what the Internet offers to support reading Terry Pratchett's *Discworld* books. In early 2007, available sites include a long-running MUD (Multi-User Dungeon, a text-based virtual world, located at http://discworld.atuin.net/lpc/, retrieved March 6, 2007) at one end of the creative spectrum, and, at the other end, detailed instructions for creating a complex Discworld wedding cake, perhaps an extreme example of expanding a fictional universe into real life (http://homepages.tesco.net/~janefisk/discworld/discworld.htm, retrieved March 6, 2007). Other sites include background information about the Discworld in which the stories take place, about the author, about other readers. They frequently include playful and expansive big-world supplements to the reading experience, such as The Unseen University Challenge (http://www.lspace.org/games/uuc.html, retrieved March 28, 2007). The 'Discworld Monthly' (http://www.discworldmonthly.co.uk/, retrieved March 28, 2007) is an online newsletter that keeps fans up to date with the newest doings in and around this fantasy universe. The print books of this series alone comprise a big world in their own right (as of 2006, there are 36 titles, according to Wikipedia [http://en.wikipedia.org/wiki/Discworld, retrieved March 28, 2007]), but the ways in which that already enormous world has been expanded by devotees is a highly instructive phenomenon that would itself repay further research. Many of these websites both manifest and scaffold a detailed interest in a particular fiction that may at times border on the obsessive, and the phenomenon raises questions about whether big worlds have any limits in the age of the Internet.

Becoming an expert in a particular fiction is one of the rewards of thick play (though thick play, in and of itself, is not confined to fiction). The idea of amateur researchers whose special interest and area of expertise is an entirely fictional field is now a commonplace rather than a faintly oxymoronic oddity. Repeated visits to different corners of this universe (through encounters with sequels and series titles, adaptations, new versions, and so forth) allow readers to accrue expertise in such components of the fiction as the history and back story of the characters and their relationships to each other, the detailed geography and appearance of the setting, and the ways in which different presenters of the story choose to highlight particular aspects of the story. For many readers, such accumulation of fictional 'facts' about the universe in which their story is set is a source of significant satisfaction. Fictional-world research can be every bit as engaging as pursuing information about real-life hobbies and interests, although we do not always recognize it as research.

Where do we draw the line between activity like Drew's detailed enthusiasm for *The Lord of the Rings* and its descendants (which might be dismissively classified as a hobby or an obsession) and the work we respectfully label as scholarship? Certainly the scaffolding to support the development of real expertise in aspects of popular culture is widely accessible and a highly regarded element of engagement with many texts. What critical components distinguish the approach of the literary historian of Tolkien's work from Drew's attentive and detailed enjoyment? Who gets to decide what detailed knowledge ranks as trivia and what gets described as depth of understanding? In a digital world where amateur experts created their own communities of interpreters, such issues of cultural capital (Bourdieu, 1979/1984) are significant.

IMPLICIT EXPERTISE

The explicit pursuit of further knowledge about a fictional world is only one way of becoming expert. A kind of implicit expertise may develop as a consequence of encountering different instantiations of the same story. I asked a group of young people aged between 7 and 14 to look at a variety of versions of *The Secret Garden* (in print, film, audio and CD-ROM format). Moving between different incarnations of the same story prompted them to comment on many aspects of the production process and many contrasts in how each adaptation achieved its effects. Looking at five different film and television versions, for example, they articulated an awareness— unprompted by me—of the differing roles of background music, which they marked even in its absence.

> Jennifer (aged 11, on the 1993 Warner Bros. film): The background music was kind of sad and in some places it was just like when she was unhappy or got a little bit happier, then there was sad music. (Mackey, 1998, p. 24)
>
> Cal (aged 14, on the 1987 Hallmark made-for-TV movie): The music that played along was an eerie music, and that played along with the lightning and the crashing; that was good. (Mackey, 1998, p. 25)
>
> Ted (aged 11, on the 1975 BBC TV serialization, which has no background music): [The music] gives it a sort of *je ne sais quoi*—sets the mood for the story. I don't really miss it but it would be better if it was there. (Mackey, 1998, p. 25)

Music, as the comments of these young people make clear, works at the border between technical considerations and affective involvement in the story. Moving between one adaptation and another of a particular story highlights the *composed* nature of that story for these young viewers and makes it easier to pay attention to different components that are used to build a recognizable atmosphere and story world.

'SLIPPERY' TEXTS

Some thick play takes place in fixed textual worlds, revisited through reiteration, as with Candace's repeated readings of *The Secret Garden*, where the sameness of the words is part of the appeal of the experience. Much textual play exists under much more fluid conditions. Writing in a different context, novelist Nora Kelly describes the transient mutability and incremental stability of daily life in terms that can be usefully applied to many forms of contemporary thick play:

> She made the turn on to Salt Hill, towards home. . . . She drove without thinking, the map printed behind her eyes. She knew every dip and rise, every pothole and frost-heave. The road was the same as always. The same. What did that mean? That she had travelled the narrow country way a thousand times, and each time she'd seen a different play of light on leaves, unique shapes of winter snow. Only by knowing these differences could the sameness—the sameness that was home—be recognized. Image was laid over image until the many layers of memory fused, became a knowledge that was in the body. She knew this road. The knowledge worked without thought, beating steadily, like her heart. (1999, p. 196)

Much contemporary thick play involves what Kristie Fleckenstein refers to as 'slippery texts' (2003, p. 105), materials that blur in terms of subject matter, genre and medium. Experience in this world is built up incrementally.

> A startling realization of this cybernetic age is that texts dissolve. Yet we are trained by our traditions and our print history to perceive our textual reality as a material one, reified in the undeniable weight and cost of textbooks. We are trained in the epistemology of presence because of these material tools, so says N. Katherine Hayles. But that is a specious stability. A text remains lodged within a single medium only temporally and sometimes not even then. All we need to do is look at the Pokémon craze, which began as a video arcade game, morphed to cards and board games, traveled from there to television cartoons, then slid over to films, cyberspace Web sites, and back to video games, this time as Game Boy games. We witness daily the dissolution and evolution of texts across media. (2003, p. 111)

It is not too difficult to see Fleckenstein's *Pokémon* players building up a big world context for their play in ways that are already familiar to them from daily life, just as Kelly describes the driver who builds up a sense of the stability of the road's contours through repeated exposure to small differences. Knowing many instantiations of a story shapes an interpreter's sense of what components of the story are crucial to a successful telling and what can be altered or nuanced. Such interpreters will vary in the priority they place on complete conformity to the original text, and in their tolerance for shape-shifting. Again, young readers will accrue a large tacit understanding of these issues, available for reference in critical discussions about fidelity and interpretation.

The question of which components are essential to a story told and retold in different media is one that taps into a great store of implicit knowledge. What liberties are tolerable, or even necessary to tell the story in a new medium? What changes are simply not acceptable, and why? Is there agreement among a group of people on the answers to this question, and what are the implications of such agreement or lack of it?

MAKING A BIGGER WORLD

Some readers and viewers express their implicit expertise in a particular story in the form of fan fiction, in which they not only recreate and elaborate on the characters and events of a favourite story but also mimic the spirit of the telling and the tone of the narrating voice. Real aficionados then may develop further proficiency in analyzing the relationship between the canonical fiction and the fanfic embellishments.

Fan fiction is not new, though it has not always borne this label. David Brewer, in a book called *The Afterlife of Character, 1726–1825*, talks about the development in the 1720s of a new reading practice that he calls 'imaginative expansion', which could take the 'liminal features of a text' and turn them into 'an opportunity for communal play' (2005, p. 26). He cites 'the unprecedented proliferation' (2005, p. 26), starting in 1726, of readers' creations which presented the story of Lemuel Gulliver outside the confines of *Gulliver's Travels* and contributed to the development of a virtual community of what we today would call fans. The instinct to create ways to linger in a fictional world that attracts our interest and affection is clearly deep-rooted. The potential for such a territory to be exciting yet safe is highly significant.

The Internet provides a natural—and very expansive—home for all kinds of forms of imaginative extension. Written stories that enlarge on a published fiction proliferate online, and it is also possible to find fan forms of film, music, gameplay and a variety of visual tributes. Many of these productions are pedestrian but their very existence is evidence of the potency of our commitment to particular stories.

It is not surprising that fan fiction is sometimes devoted to productions that in themselves are already reworkings. For example, 'How Complete a Loss Can Be' (GiaKohana, http://www.ficwad.com/viewstory.php?sid=7089, retrieved March 12, 2007) is a fan story based on the Norman/Simon musical version of *The Secret Garden* that I discussed above. It is a nonlinear story, told in extended flashbacks, but, interestingly, there is not a ghost in sight. In this case, the priorities of the fanfic writer do not include the pressures and insights that led to the 'misreading' entailed in the musical retelling.

Fan fiction resembles certain kinds of classroom activity in a variety of ways. Retelling or telling 'more' of a given story is a familiar assignment for English students. Like school assignments, fan fiction is also evaluated. Other readers respond to the story—often only to praise in the most glowing terms, but critical commentary is not unusual and the online community has built a complex set of standards about what is acceptable writing in a particular story world.

WHY IT MATTERS

Many teachers would be happy to apply the label 'exciting yet safe' to their classrooms, yet my ethnographic work with text users of various ages has outlined a form of textual engagement that is in many ways inimical to classroom work. The time pressures built into most curricula determine that leisurely, repeated and/or serial exposure to a particular textual world in many adapted forms is not a realistic option for many teachers. At best, classrooms may find ways to offer snapshot versions, small-scale simulacra of the full-throated commitment to an extensive fiction.

Yet there are many useful insights to be gained from these users of texts in many different media. Absorbed commitment to a text or family of texts provides many routes to tacit learning, and it is important to register that some elements of expertise in text interpretation are best or even only gained tacitly. Fluency, an underrated element of competence in all forms of text use, is often best achieved when attention is engrossed with content, and a territory that is exciting yet safe offers fertile ground for developing and improving assured facility with forms and formats.

I believe it is important for teachers to acknowledge and respect aspects of fictional expertise gained through play, thick or otherwise. But it is also important to consider how facility with one particular fictional world may be usefully disembedded and made available for general use. Some of the readers I have talked with over the years are committed to reading inside only one big world. Do we label these readers as specialist or intensive, or do we turn to the more judgmental adjective and call them narrow? What are the implications of this kind of restricted reading for the resources that students bring into their classrooms? If their repertoire of what makes for enjoyable reading is highly limited, if they can enjoy only what they already feel safe with, if their capacity to reject alternatives is highly developed, what approaches offer useful ways of opening them up to the kinds of varied reading done in school?

Similarly, when we look at the multiple ways in which many young readers joyously expand a favourite fictional world, we see many forms of school-like behaviour. Fan fiction involves writing around known texts and it invites elaborate critique. How do teachers acknowledge the expertise gained in such forums, while finding ways to stretch and refine responses to school-authorized materials?

It is too easy to say that the 'aesthetic of unfinish' largely relates to highly marketed blockbuster movies and their commercial spin-offs. In fact, such an aesthetic applies to all contemporary reading. Series reading is a significant component of many people's leisure choices. Many films and television programs are sequential in nature. Digital games often proliferate in series formats. Adaptations cross all media boundaries. And the Internet, perhaps the ultimate 'big world' of our times, provides expansion materials that connect to very many contemporary and classic fictions. Reading, on the page and on the screen, is a substantial part of many big world experiences, but the idea of thick play is manifest in every form of media currently available.

Not every reader, viewer or game, of course, is actually interested in this type of big world; some readers like each novel they read to live up to its nomenclature and provide a virtual experience that is genuinely *new*. But it is possible to make an argument that, regardless of whether it involves big worlds or not, thick play in itself has a role to perform in the development of committed interpreters of fictional texts. It is certainly arguable that every reader can benefit from finding a connection to a print text (fiction or nonfiction) that leads to some form of thick reading play.

Stephen Krashen makes the detailed case for free voluntary reading (FVR) as an necessary component in the development of reading skills:

> FVR is one of the most powerful tools we have in language education. . . . FVR is the missing ingredient. . . . It will not, by itself, produce the highest levels of competence; rather, it provides a foundation so that higher levels of proficiency may be reached. When FVR is missing, these advanced levels are extremely difficult to attain. (2004, p. 1)

Such leisure reading is a major form of thick play, and Krashen's work demonstrates that it is an indispensable element of becoming a skilled and confident reader. Reading only for school assignments simply does not provide sufficient bulk opportunities for practice; playful reading time is also important.

Similarly, computer skills thrive in a setting of thick play. As many task-oriented adults know to their cost, the teenagers who spend hours just fooling around with digital machinery, in various playful and time-consuming ways, usually wind up knowing much more about computers, and feeling at home in a plethora of digital worlds.

In both cases, explicit knowledge is accessible through direct training—but implicit understandings that enrich and enable thick encounters with a medium are most successfully built up thickly, through repeat playful encounters. This statement, which seems so self-evident as to be tautological when expressed in these bald terms, is clearly not so obvious in the real world, where faith continues to flourish that phonics instruction and computer training courses are sufficient and successful routes to fluency and enjoyment.

Finding ways to sponsor thick play in the literature classroom may not be simple, but opening the eyes of teachers and students alike (as well as parents) to the virtues of thick reading play elsewhere is a valuable exercise. Relishing a particular big world is certainly one organic route to the acquisition of the thick play habit.

CONCLUSIONS

Thick play and big worlds are concepts that apply across many different media. In this chapter I have made a point of mixing examples of print reading with experiences in other media; too often teachers and researchers are inclined to ring-fence reading into its own special territory. Outside of school, young people read as part of their encounters with story in a broad range of formats; we do nobody a favour by marking out the activity of reading as some kind of precious exclusion to generalizations about popular culture.

As well as drawing on a pool of different media references, I have also explored the activities of a range of ages. Contemporary young people look for role models of literate adulthood to those adults who are closest to them in age, as the frontline representatives of what it means for full literacy to grow up digital. Researchers and teachers can also learn a great deal from these newest adults.

Finally, I have explored the significance of leisure activities with a view to gleaning insights that may be useful for teachers. We live in a time of profound disconnect between in-school and out-of-school literacies. Classrooms may not be able to replicate the conditions that enable thick play and big-world exploration in all their full, unhurried ramifications. Nevertheless, there is much to learn about the significance of specialist expertise and comfort, and the conditions for maximum development of tacit skills—both in print and in old and new media. True excitement combined with true safety is a powerful support for learning.

REFERENCES

Bixler, P. (1994). *The Secret Garden* 'misread': The Broadway musical as creative interpretation. *Children's Literature 22*, Journal of the Modern Language

Association Division of Children's Literature and of the Children's Literature Association. Yale University Press, 101–123.

Bourdieu, P. (1984). *Distinction: A social critique of the judgement of taste*. (R. Nice, Trans.). Cambridge, MA: Harvard University Press.

Brewer, D. A. (2005). *The afterlife of character, 1726–1825*. Philadelphia: University of Pennsylvania Press.

Csikszentmihalyi, M. (1997). *Finding flow: The psychology of engagement with everyday life*. New York: Basic Books.

Fleckenstein, K. S. (2003). *Embodied literacies: Imageword and a poetics of teaching*. Carbondale, Il: Southern Illinois University Press.

Geertz, C. (1975). *The interpretation of cultures: Selected essays*. London: Hutchinson.

Hutcheon, L. (2006). *A theory of adaptation*. New York: Routledge.

Kelly, N. (1999). *Old wounds*. London: HarperCollins.

Klinger, B. (2006). *Beyond the multiplex: Cinema, new technologies, and the home*. Berkeley, CA: University of California Press.

Krashen, S. D. (2004). *The power of reading: Insights from the research* (2nd ed). Westport, CT: Libraries Unlimited/Portsmouth, NH: Heinemann.

Lunenfeld, P. (2000). Unfinished business. In P. Lunenfeld (Ed.), *The digital dialectic: New essays on new media* (pp. 7–22). Cambridge, MA: MIT Press.

Mackey, M. (1995). *Imagining with words: The temporal processes of reading fiction*. Unpublished doctoral dissertation. Edmonton, Canada: University of Alberta.

Mackey, M. (1998, Winter). Tacit literacies: The challenge of exploring representations. *Alberta English, 36*(1), 22–28.

Mackey, M. (2006, June). Serial monogamy: Extended fictions and the television revolution. *Children's Literature in Education, 37*(2), 149–161.

Mackey, M. (2007). *Mapping recreational literacies: Contemporary adults at play*. New York: Peter Lang.

Maynes, W., & Mackey, M. (2006, Spring). Narrative attraction: One story, one reader, and two decades of ongoing appeal. *Language & Literacy, 8*(2). Retrieved February 11, 2007, from http://www.langandlit.ualberta.ca/Spring2006/Maynes.htm

Schoen, B. (2006). *Mugglenet.com's what will happen in Harry Potter 7: Who lives, who dies, who falls in love and how will the adventure finally end?* Berkeley, CA: Ulysses Press.

Thompson, C. (2007, September). The science of play. *Wired, 15*(09), 140–147, 184, 192.

WEBSITES LISTED

http://en.wikipedia.org/wiki/Halo_3#Cooperative_play, retrieved August 26, 2007
http://discworld.atuin.net/lpc/, retrieved March 6, 2007
http://homepages.tesco.net/~janefi sk/discworld/discworld.htm, retrieved March 6, 2007
http://www.1space.org/games/uuc.html, retrieved March 28, 2007
http://www.discworldmonthly.co.uk/, retrieved March 28, 2007
http://en.wikipedia.org/wiki/Discworld, retrieved March 28, 2007
GiaKohana, http://www.fi.cwad.com/viewstory/php?sid=7089, retrieved March 12, 2007

7 Online Connections, Collaborations, Chronicles and Crossings

Julia Davies

INTRODUCTION

Rhys, an English boy in his mid-teens, sits with his parents in a hotel lounge in Berlin. Each evening of my summer holiday, I see him use his laptop computer while his parents relax with a glass of wine; there is a sense of the habitual about this arrangement, the parents talking together and the son inhabiting a virtual space with his friends. Rhys begins each evening ritually opening several windows at once, swiftly moving through them—clicking, sliding and tapping on pads and keys. He looks only at the screen as he accesses 'MySpace', e-mail, Instant Messaging, YouTube and other sites as suggested to him by friends during his online interactions. 'Pop up' ads sporadically shoot into view, and zapping them like an automaton, he expertly clicks them away as suddenly as they appear. He is happy to chat to me at the same time and explains he is 'with' school friends, just as he is most nights at home. He passes on jokes and makes them; they collaborate in commenting on digital photographs he has taken in Berlin and make up titles for them; he tells me that when you are online you can 'do stuff with your mates' and just 'hang out'. His face seems animated as he takes part in a range of interactions (two of his friends are on holiday and one connects (albeit intermittently) through a mobile phone, while another uses her home pc) and it is as if Rhys occupies more than one space, keeping up a kind of polysemic identity, so that while remaining in the familial holiday context he maintains an online presence which blends his holiday, school and home selves.

It seemed to me that Rhys was not merely 'keeping up to date', or 'catching up', but contributing to 'happenings' online. This was more about 'participation in online events' than 'keeping in touch'; that is, Rhys and his friends were involved in collaboratively producing texts which constituted social moves which had the potential to affect their sense of identity and their relationships with each other. They were involved in multimodal text production; they dealt in still and moving images; written words; sound and emoticons. Drawing on observations from a wider sphere, and as outlined in this chapter, I would argue that Rhys's relationships with his friends were extended and constructed through their playful multimodal digital text making.

PLAYFUL SPACES

Like many others I have observed over years of research in this area, Rhys and his friends played with words—their sounds and their meanings; with images—their meanings, appearance and so on as part of their exploration of the potential of the web, of themselves and of ideas. I have seen people involved in online sociodramatic play (Davies, 2006b; 2006c, 2006d, 2007). They implicate themselves within the texts, so that they become 'inter-textually constituted'—that is to say their identity is made up partly through how they present themselves online and what they do there, and of course partly through their knowledge of each other in offline spaces too. Not only this, because participation occurs online, they can challenge barriers of time and space—they can be together even when apart and can contribute at different times to one another—or if they wish—at the same time. This blurring of boundaries, of the online and offline self; of online and offline activity, is part of the allure of text-making and constitutes the social world of more and more people in our society.

In previous work (Davies 2004; 2006d) I described how girls' online play with *'Babyz software'*, (Mindscape, 1999) allowed players to broaden the scope of their world and to get to know new people online through their play. The Babyz software was marketed strongly towards the pre-teen (or tweenie) girl market and provided a playspace for traditional doll playing with the bonus that the virtual doll could move, talk, eat and be seen to develop, albeit in a virtual world. In terms of values, the game had everything in common with traditional mother and baby play. Many of the girls who played this game also moved their activities into the online arena and were able to interact with other Babyz players. They collaborated over the design of new baby clothes, new 'toyz' for 'Babyz' and even created play scenarios such as virtual schools and hospitals. The 'Babyz Community' ('BC') (as they referred to themselves), used the affordances of different sites for different purposes and they collaboratively adapted the text that constituted the game—the images, words and music, etc. They sent images to each other, even gave each other passwords to sites they created and shared in the design of new online spaces. Their collaborations were focused on play and connectivity.

These new digital practices are not, as some may argue, confined just to the young; indeed, research suggests that in the United Kingdom at least, women between the ages of 18–34 are now the most dominant online group (Nielson/Net Ratings, 2007), involving themselves in online social networking on a daily basis. The 'digital generation' is as Buckingham (2006, p. 2) argues, perhaps just 'easy rhetoric'. Participants in this newly and still emergent digital era are a diverse and disparate population who are embedding technologies into their lives as a lifestyle choice for many different reasons—they cannot be identified as belonging to a single age group. Links are being made across social and geographical spaces and time zones; it seems to me that the 'digital divide' is less simple than being age-defined and there are many kinds of digital native (Prensky, 2001) involved in many kinds of activity. Connectivity is increasingly part of adults, working and social lives; like the young, they use the Internet as a way of exploring new ways of connecting socially and the skills to combine these areas seem to be increasingly valued in both the private and the public spheres. Presenting the self in different ways within and across a variety of online spaces, as well as linking with likeminded others can be done very easily by using a range of textual devices. In this chapter, I argue that online, textual cohesive devices afford increased opportunities to link socially. In this chapter, I show how this can be done, drawing on observations taken across the age range, from tweenies to adults and in particular focus on playful uses of technology as a way of exploring the self, the world and our place within it.

COLLABORATIVE NARRATIVES AND WRITINGS OF THE SELF

Over the years, researching aspects of online activities, my own involvement has become increasingly participatory. Initially I stood 'outside', lurking on websites and discussion forums, but like Markham (1998) have found an insider perspective useful. In a determination to better understand, I began keeping a blog (a 'web-log'—a regularly updated website, organised in date order), and found myself quickly drawn into a network of associated online activities. This process has allowed me to build relationships with others online and experience the overlap of online activities with other aspects of my personal, social and cultural life. When I first started my blog, it was not possible to upload images directly to it and so at first I had to upload to a photo-sharing website, Flickr.com. (This allowed me to get [html] code for my images which I could then paste into my blog post in order to use that image online). I was surprised to discover a whole new community of online enthusiasts who were eager to interact and discuss images—it seemed that for many, the site was more about interactivity than just a space to store and view images. Just as I had observed with the much younger Babyz Community (BC), members were keen to invest in the affordances available, even, as I shall show, pushing the boundaries of what was available in order to be creative. My research showed that in the BC, play was often about being connected, rather than about the game itself; it was about participation and making a mark in online spaces—for example, some of those who participated in the online Babyz play did not even possess the game. What I witnessed with Rhys in Berlin, with his perpetual use of technology to maintain connectivity, and what I continue to witness in so many locations (the university library, the railway station, airports, Starbucks cafes and even in parks), exemplifies the ways in which new technologies are developing and sustaining new ways of living and being as well as providing instantaneous and 'always on' networking.

In this chapter then, I draw upon observations of my own practices, as well as upon my observation of others over the years. The data from which I have been working includes social networking sites like Flickr.com, Facebook.com, a range of blogs, as well as a number of teenagers' personal websites and a number of discussion boards. I have adopted a range of analytical approaches, but in particular have highlighted interactants' abilities to learn through informally organised activities. In this chapter, I particularly highlight the ways in which participants exploit the internet's social networking affordances. I discuss some of the cohesive features of digital texts and explore how playful online interactivity exploits these cohesive features in order to keep 'networked'.

NEW LITERACIES

My work is situated within the 'New Literacy Studies' (NLS; Gee, 1996; Street, 2003). The NLS is a new tradition, emphasising literacy as a social practice, rather than—as in the 'old' tradition—being purely skills-based. The pluralisation of the term 'literacy', in the 'New Literacy Studies,' recognises that literacy practices, *because* they are social, vary according to social context and that literacy is not monolithic.

Like other researchers looking at digital text-making practices, (or *'digital literacies'*) Knobel and Lankshear (2007) have emphasised the ways in which 'New Literacies' cannot be merely defined by their means of production via 'technological stuff' (digitality). They argue that it is not gadgetry alone that defines 'New Literacies'; and they refer to 'new ethos stuff' to explain. This definition of New Literacies takes into account the *social uses* of digital text-making practices, so the paradigm of the New Literacy Studies makes an ideal lens through which to explore New Literacies. In describing the new kind of 'ethos stuff', Knobel and Lankshear (2007, p. 9) make the distinction from older styles of literacy, saying it is more collaborative than 'individuated'; more 'participatory' than 'author-centric'; less 'published', than 'participatory'. 'New Literacies' are therefore often to be found online where textual and social cohesion have a synergy where they each affect the other. Textual cohesion is an important focus for this chapter and it is often through textual cohesive devices that play is afforded; cohesive devices allow connectivity not just across texts, but also across people.

This chapter describes some of the new ways of behaving and connecting using new technology; it explores how the modes and the materiality of digitally produced texts, facilitate the development of a new ethos in text and meaning-making for its users. Further it shows the ways in which textual and semantic cohesion facilitated by technology can produce collaborative, participatory texts which support individuals to communicate in innovative ways. Moreover, in line with the theme of this book, the chapter focuses on playful approaches in text production and consumption and how this is conducive to learning.

BLENDING SPACES AND DEVELOPING AFFINITIES

There are many ways in which people can make links with each other in online spaces; one way is to continue conversations begun in 'real time' in face-to-face situations. Benkler (2006) argues how online interactivity 'thickens' existing social ties and boyd posits:

> The digital era has allowed us to cross space and time, engage with people in a far-off time zone as though they were just next door. . . . (and) . . . network us all closer and closer every day. Yet, people don't live in a global world—they are more concerned with the cultures in which they participate. (boyd, 2006, p. 1)

There is a strong sense here then, that the social aspects of the web are being seen as a way of continuing existing relationships, of providing an additional dimension to play out friendships and an arena where important socialising can take place. For many web users, the boundaries between offline and online worlds and activities are blurred and it is clear that many MySpace and Facebook users mainly, or exclusively, communicate with others also known to them offline; much of their relationship management occurs in online spaces (boyd, 2006; Dowdall, 2006). Online, people can discuss and reconfigure their local lives in a broader cultural context (Davies, 2004, 2006a; Davies & Merchant, 2007; Lankshear & Knobel, 2006); they can view themselves 'out there', or see the local on a global stage. As Bortree shows (2005), bloggers represent their offline lives, whilst also creating an online identity in the 'blogosphere' through their textual practices; outcomes of these practices can then emerge in offline lives and back again. I have seen how individuals can come together online and develop shared practices and ways of behaving and spend so much time together that a kind of social history develops through shared multiply authored events or in-jokes (Davies, 2006a, 2006c, 2007) and all these representations of online selves can seep into offline lives and vice versa.

MULTIMODALITY

Cohesive devices are the links that hold a text together and give it meaning. They need not be linguistic and can be semantic, grammatical, aural or visual. Using semiotic and discourse analysis techniques, (Kress & Van Leeuwen, 1996) I look at the ways in which digital texts work *as* texts, and at how specific textual features are used to perform identity and at how they work socially. That is, I consider how hypertext, images, digital video and sound, as well as written words, can be used to link smaller texts authored by a range of individuals, to function as a larger more complex multiply-authored, polysemic, multilayered text. I argue that through the production of texts which link with each, individuals can collaborate and weave a network of associations which are both explicitly articulated and also implicitly expressed through a variety of devices which make texts cohesive. The cohesion works on more than one level, being related both to the *structure* (e.g., the grammar or layout) of the texts, the *modes* (e.g., images or written words*)* and also to the *semantics (or meanings)*. The ways the texts are composed and the way they cohere, allows particular kinds of textual self to be presented.

I consider the methods and textual devices individuals use to form or deepen social allegiances online; I look at how they share and develop meanings, create social histories which help define specific groups and use the web primarily as a place in which to socialize.

HYPERTEXT LINKS AND PUBLIC DISPLAYS OF CONNECTION

Hypertextual linking is perhaps the most distinctive feature of online texts; it allows writers to add code to their work which allows navigation across or within texts. This basic tool means that many texts can be joined or embedded within each other and where links are reciprocal, textual relationships can be multiply-layered. This embedding allows a kind of 'ventriloquism' or 'double voicing' (Bakhtin, 1981) to occur where a writer can easily incorporate the voices of others in their own narrative (Davies & Merchant, 2007).

Through hypertext other cohesive features have developed, such as the blogroll, which allows one blogger to list and provide a hypertext link to, other blogs. This is social networking behaviour at its most basic, (Danath & boyd, 2004, p. 2) with public displays of connection being key to a demonstration of a particular type of identity. Through the display of affiliations one chooses to (or not to) reveal the types of alliance as well as the number of alliances made; these can form part of one's online identity.

Below we see how this can be worked in a playful way:

Stolen from *Skits*, who stole it from *Meeta*, who stole it from *Busy Mom*, who stole it from *Jill*, who stole it from *Neva Miss Feva*, who stole it from *3rd House Party*, who stole it from *Twilight Cafe*, who stole it from *Pam*, who stole it from *Jo*, who stole it from *Whump*, who got it from *tamiam*, and oh, I give up.

Just do it (which I blatantly steal from Nike).

(bozbozboz, 2004)

In quite an extreme demonstration of connectivity, shown in this extended sentence bozbozboz acknowledges the source of an idea he is using in his blog. There is a definite ironic tone here, as bozbozboz acknowledges all those who have influenced him; he carefully shows how in attributing acknowledgement online, it can be a long complex process since so many texts are embedded within each other. The embeddedness of the subordinated clauses grammatically enacts the embedded nature of the idea as it has passed through the Internet from person to person, text to text. It is evocative of the children's nursery rhyme, a nonsensical verse which talks about 'The house that Jack built'. The blogger affiliates his text in this way to children's culture as well as to other bloggers in an elaborate demonstration of how he is very well networked. His use of grammar with a series of subordinated clauses, along with hypertext, is playful and emphasizes his online connectivity. He also illustrates how it is possible to trace ideas online. Each italicized word on this text represents where a hyperlink lies; one can link directly from this text to all those other bloggers this blogger has acknowledged. In this way the texts of those blogs become incorporated into bozbozboz's blog, and simultaneously they are drawn into a delineated affinity, where they are linked textually. We see how the blogging practices of many bloggers can contribute to one specific literacy event. Thus in outlining the provenance of his own work, bozbozboz reflects on ways in which the narrative of his text links to other texts both structurally and semiotically.

MEMES

Dawkins (2006) used the term 'meme' as a metaphor to describe the process through which ideas spread and develop through cultures. Memes are analogous to genes, in that for a particular idea to be sustained and passed on, a particular culture must be predisposed to accept it; mutation/adaptation will occur in strong memes in order for it to survive in different cultures. 'Selection favours memes that exploit their cultural environment to their own advantage' explains Dawkins (2006, p. 199). Thinking broadly, Dawkins gives examples of memes as:

. . . tunes, ideas, catch-phrases, clothes fashions, ways of making pots or of building arches. Just as genes propagate themselves in the gene

pool by leaping from body to body via sperms or eggs, so memes propagate themselves in the meme pool by leaping from brain to brain via a
process which, in the broad sense, can be called imitation. (Dawkins,
2006, p. 192)

Digital technology makes the duplication of text, and replication of
sound and images easy and so can be quickly copied, developed and trans-
mitted across vast geographical, social and cultural areas. On the Internet,
where connectivity is easy, the cross-fertilisation of ideas is rife and it is
often difficult to see where a particular idea began, because of the constant weaving of texts, in and out of each other. Memes can be passed on
through hyperlinks; through narratives that have meshed together; through
online jokes; through images passed from screen to screen and copied or
altered in some purposeful or accidental way; Knobel and Lankshear have
discussed a number of well-known memes (2005) which behave in this way.
In my observations, online individuals seek connectivity; they *make* rather
than break chains, so memes can become rife, and of course the example
from bozbozboz above, is a meme.

On Flickr.com, a photo-sharing website, the use of old cameras alongside digital technology has spread like a meme. It has become popular to
upload images taken by two cameras together—with the lens of one camera
looking through another. Usually the second camera is antiquated and non-
digital so that resultant images are blurred, dirty and seemingly antiquated.
The images are often only partly in frame and appear less sharp than conventional digital images (BurpsLiberty, 2006). Many examples have been
posted to a group called 'through the viewfinder' and it is clear from online
discussions within the group (through the viewfinder, 2006) that the play
with the images takes place across a range of online and offline spaces;
participants buy old cameras from eBay; some meet together to show each
other their cameras and techniques; they repeatedly experiment with shots
and processing and have discussions about individual images online. The
play is ironic in its use of digital technology but also seriously explores the
potential of technologies and considers how different kinds of processing
affect meanings within images. Here, we see an example of how online play
involves a number of spaces, objects and modes and media, exploring different ways of representing the 'real' world within the online space.

I mentioned earlier the Babyz Community (BC) and my observations of
the ways in which they developed the game to suit their own ends—designing schools and clothes and exchanging them amongst themselves—beyond
the parameters set up by the game's designers. Many BC players linked to
each other's sites, developed shared discourses for describing their play, and
passed on tips for how to manipulate computer codes to achieve specific
looks for Babyz and their clothes. Such interactivity reflected shared values
and could be seen also as 'memetic' behaviour. Further, some of the BC sub-
verted the stereotypical (sexist) values of the game, even manipulating the

digital code to change the appearance of the Babyz. They moved beyond the official online space of the game, subverting assumptions about how they should 'play nicely' and refused to play like stereotypical loving mothers. They turned their babies into hairy monsters for example and sometimes stopped feeding them. In order to do so they needed to become involved in altering the way the Babyz were designed, manipulating computer code and so on. Within the self-named 'Babyz Community' or 'BC' this was a small but much noticed group of girls who had developed their online play in this way; those who participated in this play aligned themselves closely and shared an affinity through 'aberrant' behaviour. They even set up a site called 'Abusing Babyz.com' and were quickly made aware by others in the BC that this was offensive to many of them. On other sites girls were making anti 'abusing babyz' banners for websites and made these freely available for digital copying. Similarly, these girls were also showing they shared values and 'meme-like', their banners proliferated quickly and allegiances were swiftly decided. Debate was rife about whether purposefully not feeding a cyber baby indicated a likelihood of real child abuse in later life; knowledge of the debate indicated strong membership of the BC and participation was to take part in the meme.

At the time of writing, a popular meme on Facebook.com, (a popular networking site) is to cast a Harry Potter spell on a Facebook friend. This draws on the current heightened popularity of the infamous series of books by J. K. Rowling and coincides with the release of a new film. The intertextual linking to a popular film that appeals to adults and to children means that the meme can pass quickly amongst members of Facebook—happening by means of one individual leaving a message on someone's Facebook, saying that a spell has been cast on them. One of a selection of icons signals that the spell has been left and the recipient is given the option of continuing the game by casting a spell on someone else. This is of course symbolic play, and only text—words and images—constitute a 'spell' being conjured. This kind of meme is very common in the Facebook space and helps individuals to demonstrate an online sociability, even when there is no actual conversation (in the usual sense) happening. As with the last example, participation in the meme requires cultural knowledge gained within and beyond the site, in both on- and offline spaces. It is clear that the spells are just online play; no one believes that a spell has been cast, but the ritual is playful and is about affirming friendship and knowledge of particular cultural and social happenings.

A meme in which I became particularly interested was rife across blogs around 2004–2006 and could be described as the 'fake memory of me' meme. The meme was of interest because of its play with identity and semantics. Individuals invited others to invent a memory of them; in inventing a story, the text became part of the original blogger's textual self, so that it formed part of an online social history, even if the event did not occur in reality. It is about the self in this context, an online self, which may or may not carry over into other contexts. Here bozbozboz describes how she came across the meme, and joins in with the game by inventing a story about herself:

Reinventing history

Here's a meme from Jo to get us through the rest of this interminable week:

Invent a memory of me and post it in the comments. It can be anything you want, so long as it's something that's never happened. Then post this in your journal so that people can invent memories for you.

I'd appreciate it, though, if we could let rest the little incident with the yacht. Nobody pressed charges after my capture, and besides, it all happened in international waters—jurisdiction was so murky back then.

(bozbozboz, 2005)

This particular meme reflects something of the way in which blogs can be used to perform textual alliances; through the creation of fictitious happenings, commenters create an invented social history. This has the potential of becoming an 'in-joke' which can be developed much further and thus the narrative becomes a real social history—an online social history of textual selves in this particular space.

To understand a joke is to be part of the 'in crowd' and to link to the site of its origin, for example through one's blog, is to make the text part of one's own blog post. As we saw with bozbozboz's post above, the embedding of texts with multiple authorship, distributes narratives further and the use of hyperlinks develops the semantics within the context of each new text it is linked to. The provenance or the journey the meme had taken, gave additional meanings to the act of inventing a memory, since to join in is to participate in a project whose parameters are partly unknown. The whole meme is multiply authored and read by many.

Another meme in which I was involved, began where C-Monster (2003) posted an image of novelty cakes in a baker's shop in New York. The cakes had faces with large features and were covered in brightly coloured icing—the kind of confectionary one would see at children's parties. C-Monster named these cakes 'sugar dudes', because each cake looked like 'a cool little creature' (C-Monster, 2003) . Through this titling, C-Monster gave the cakes a 'hip', 'cool' character which others extended over a period of months through a series of responses from others (all adults). These responses included verbal comments, links to images others had taken of the cakes, or of similar ones elsewhere. Thus one person linked to images of cupcakes that looked like frogs, and another contributed an image of knitted frogs that looked like cakes (ljc@flickr, 2006).

The absurd humour, presented through images contributed by people who did not know each other, gave them an opportunity to make links in playful ways, interacting over cakes that were ostensibly made to appeal to children. It seemed that an interest in childlike objects was providing a focal point for online connectivity; the use of self-deprecating absurdist humour was bringing a group of people together—people who were local to the bakery, but also those from elsewhere in the United States, in the United Kingdom, Denmark and Sweden. Part of the joke, was the way the cakes were taken seriously and spoken about as if real; other aspects of the joke included images of the dudes involved in adult activities—such as with cigarettes or 'drinking' alcohol. Gradually 'the dudes' became an in-joke, understood by some and not by others. Those who had been interacting over a long time therefore got drawn closer in the group. This process gave a kind of social history to the group of people who were able to have fun in the online space; those involved also started talking about other things and it was clear that the online play had allowed individuals to start to trust each other because of the collaborative play.

This imaginative, ludic play led to a face to face meeting where some of those involved collaboratively produced a series of images depicting a narrative of the cakes 'escaping' from the Bakery. The photographs were put online as a story (TroisTetes, 2006) entitled 'The Great Escape'—linking inter-textually to the film of the same name in order to make an ironic joke—that the escape of the Sugar Dudes is a serious fact based reenactment. So many more images were taken than were uploaded online; like much play, some aspects were taken very seriously even though it was all an 'as if' situation. The activity was as much about performance and display as the process itself; for while the whole thing was based on an extended joke, nevertheless it created an opportunity to take a range of images that would be put on the Internet. Images included pictures of the story being set up and small props being made, (a metanarrative), as well as the story itself. The social metanarrative thus became woven into a kind of online sociohistorical text and the display was partly about demonstrating the friendships that had been made during the fun. The activity was also about text making and the development of a particular type of online textual self; further the play allowed relative strangers to build on tenuous connections which had been made online—the play was a way of making of friendship, which as Crystal (1998) argues, can cut through social barriers of profession, age and class, for example.

The movement to and from the online space for narrative development and text making, exemplifies the ways in which the digital landscape impacts in sociocultural ways beyond online boundaries. Months after the event, the activity continued to be enjoyed by a range of online commenters who continued the idea with stories of anthropomorphic cakes of their own, making sets of images on their streams in similar ways, such as the work in Sweden of Ruminatrix (2006). Such game playing, whilst on one level fun and whimsical, reflects ways in which online cultures can develop in unpredictable ways, demonstrating that text making is a social practice and that play can help perform important social work. The process, where online relationships were developed offline, where a narrative was developed through the use of images, allowed the participants to develop and extend their online identities as creative and fun, also presenting a textual display of their friendship through images shown online. The sociodramatic play that these adults were involved in helped them to break down social barriers and gave them a common place to start making new friendships upon which they were able to extend their online relationships.

These examples show ways in which narratives can cross through thresholds between online spaces as well as working through both on- and offline spaces and identities. Walker's work (2004) reflects on the ways in which hypertext and joint authorship allow narratives to work across space and time, developing the concept of 'distributed narratives'. She explains that distributed narratives are 'stories that aren't self-contained. They're stories that can't be experienced in a single session or in a single space'. (Walker, 2004, p. 1). She further explains that:

> Distributed narratives demand more from their readers than reading or suspension of disbelief. They ask to be taken up, passed on, distributed. They seek to be viral, the memes of narrative, looking for readers who will be carriers as well as interpreters. (Walker, 2004, p. 20)

The Sugar Dude narratives traversed a range of spaces and became a meme, an idea that others took up. The Internet lends itself well to this kind of play and because memes often are games which follow rules, they provide a structured 'way in' for newcomers, a 'foothold' to start off in a new group. Just like children on a playground, much online interaction develops rituals and conventions which aid interaction and which is ostensibly about one thing, but really is about friendship and connectivity. In this way we can see play as a socially cohesive device.

INSIDER VOCABULARY/JARGON

The web provides spaces where like-minded people can share interests together and within particular sites specialised language is inevitably used. This helps interactants talk in precise ways about things for which they have developed communal understandings over time, but also helps them to signal those shared understanding and values within and beyond the community—they are a badge of belonging, giving a sense of unity and exclusivity. I have previously discussed this in relation to a discussion board used by those who present themselves as teenage witches (Davies, 2006c). For example, they use the spelling 'Magick' instead of 'magic', and their usernames fit the genre, such as 'WildFyre', or 'Ravenwolf'; these are markers of the teenage online Wiccan community. As such these particular forms help the teenagers present particular online identities in role 'as witches'. Such insider vocabulary is ubiquitous across message boards—be it cyclists' forums, discussions about Endemol's 'Big Brother', or even the United Kingdom's 'The Archers' radio programme.

Language and other modalities are used to reflect social knowledge, and sometimes jokes develop over time and across multiple online spaces. For example, DrRob, a blogger who became linked to my blog and to that of other academics with whom I associate, brought together a phrase we had been playing with in a ludic way (Crystal, 1998) SimplyClare, commenting on my blog, had used the phrase 'well to the jaybad', which she later explained to Mary Plain:

> Mary Plain, I am sorry to admit that I was a school girl in leafy Epsom. I went to Wallace Fields Middle School at this time, and hung (hanged out?) out with many other preteens who used to say things like "Top of the Pops was well to the jay bad last night." Not too far at all from Clapham and Tooting in fact—but this terminology was obviously 'insider knowledge'.
> (SimplyClare, 2005)

Subsequently, the affinity of bloggers SimplyClare, Digigran, DrJoolz (me) and DrKate, used the phrase regularly amongst themselves; it was a way of joking, play-acting an aped youthfulness; being self deprecating, and enjoying a shared but relatively private joke. The schoolgirl provenance of the phrase was brought to bear in the new online context and was used in a fun way. This joke (and others) opened out our academic relationship giving it dimensions we had not explored face-to-face; as such we were playing with our academic identities. SimplyClare revealed something of her past as a 'schoolgirl in leafy Epsom' taking a risk, confiding about childhood language play. In risk-taking she demonstrates trust and 'thickens social ties' (Benkler, 2006). Crystal (1998) suggests that language play cuts across regional, social and professional divisions and I have often seen it employed online in this way. In a bid to join this affinity, DrRob, another blogger, adopted the phrase in poetry, as in this extract:

> Well to the jay-bad-unbad-goodbad
> Yakking on the myriad language launch pad
> . . .
> Listen to the lingo that makes you glad
> Word plaid, strictly rad,
> Well to the jay-bad-unbad-goodbad
> (DrRob, 2005)

The play with words echoes traditional childlike poetry and fits with how we had all been using the phrase. DrRob's adoption of our language was flattery and showed he had attentively read our blogs and understood our humour; in observing the linguistic rules and meanings for our new phrase, DrRob extended our wordplay. In the same post DrRob placed links to several other blogs to show where the phrase came from and in this way textually affiliated himself to all those bloggers. DrRob appears as a link on all the blogs within that 'constellation' of academic bloggers. This demonstrates how textual practices merge with social networking and how playful discourse can have a major role in defining affinity spaces. Moreover, while most of the participants in this online game were known to each other through an academic context, DrRob is known only through blogging and this reflects the varied ways in which online relationships can be configured. He used strong textual play to make his connections sustainable with the rest of the group. The phrase 'well to the jaybad' could be described as a 'meme', a feature which DrRob used to great effect. In this way the ludic language use (Crystal, 1998) was used to make links with others, a kind of 'rhetoric' of cohesion as I hope I have shown throughout the examples given in this chapter (p. 1).

CONCLUSION

I have not, in this chapter, offered a definition of play, for as Sutton-Smith (1997) argues, it is a slippery concept, with 'much ambiguity'. Nevertheless he argues, 'we all know what playing feels like' (Sutton-Smith, 1997, p. 1) and I have taken the liberty of depending on the reader's intuition in this respect. I have offered examples of playful activity online and in doing have attempted to reflect some of what is compelling about online interactivity. I have also shown the relationship between play and learning and the circuitous route around which learning is sometimes achieved—through experimentation, the breaking of conventions and in the examples I have offered, collaboration amongst people. They have used the resources offered by others and worked with those, building bonds and showing new possibilities.

In the example I have given about girls and their play with Babyz, I showed girls subverting the notion that all girls play nicely and how they were able to experiment with ways of resolving real conflicts. It allowed them to push the boundaries of stereotyped roles, to adopt a range of voices and try out different ways of being. As Kenway and Bullen (2002) argue, this is a unique offer for the young:

> The Internet is not only a source of information and games for students. It offers children and youth a means to 'distribute' their voices and views in ways that they enjoy. It also offers them the opportunity to blend the playful and earnest. (p. 181)

But I have also shown how adults play online too. Fernback (1997) describes the space offered by the Internet in this way:

virtual interaction gives users back some of their humanity—a human-ity which is authentically expressed among its constituents via a mass medium whose content is not wholly determined by corporate executives. (p. 46)

Whilst young people look for ways of empowering their voices for different reasons to adults, Fernback presents here some reasons why adults might want to feel the benefits of a playful space online. The ways in which I observed the Babyz Community operate almost a decade ago, where participants exchanged content for websites and where they were able to comment on, manipulate and add to each other's texts, are activities which now have become much more commonplace. These kinds of collaborations are associated with the term 'Web 2.0' (O'Reilly, 2005) where users upload content, adapt existing content, work collaboratively and form online connections. It seems that the technology has developed responsively, following habits of early adopters, and that the appeal of New Literacies is immense.

This chapter outlines how individuals use the social affordances of digital texts online. I have concentrated upon literacy practices which reflect what Lankshear and Knobel have referred to as 'new ethos' stuff (2007) and in particular have focused on ways in which cohesive devices have been exploited. The texts have to different extents drawn their meanings not just on their own terms, but also in relation to others; the meanings have been partly dependent upon being seen in relation to others. Participation in collaborative meaning making is a key skill to be acquired online if individuals are to be able to participate in a full range of cultures—which move across a range of spaces within and beyond digital texts.

REFERENCES

Bakhtin, M. (1981). Discourse in the novel. In M. Holoquist (Ed.), *The dialogic imagination* (C. Emerson & M. Holquist, Trans.). Austin, Texas: University of Texas. pp. 259–622.

Benkler,Y. (2006). *The wealth of learning networks*. E-book—Chapter 10. Retrieved 30th November 2006, from http://www.benkler.org/Benkler_Wealth_Of_Networks_Chapter_10.pdf

Bortree, D. S. (2005). Presentation of self on the web: An ethnographic study of teen-age girls' weblogs. *Education, Communication and Information, 5*(1), 25–39.

boyd, d. (2006). G/*localization: When global information and local interaction collide.* O'Reilly Paper given at the Emerging Technology Conference, March 6, 2006 . Retrieved August 2006, from http://www.danah.org/papers/Etech2006.html

bozbozboz. (2004). Remember when. [Blogpost]. Retrieved September 2006. (Web address withheld on request.)

Buckingham, D. (2006). Is there a digital generation? In Buckingham, D. and Willett, R. (Eds.) *Digital generation: Children, young people, and new media* (pp. 1–18*)*. London and New York: Lawrence Erlbaum Associates.

BurpsLiberty. (2006). Papa Burps. An image on Flickr.com within the group 'through the viewfinder.' Retrieved August 1, 2007, from http://www.flickr.com/photos/burpsliberty/323679038/

C-Monster. (2003). Sugar Dudes. An image on Flickr.com. Retrieved 30th November 2007, from http://www.flickr.com/photos/arte/4424203/in/set-112574/

Crystal, D. (1998). *Language play*. London: Penguin.

Danath, J., & boyd, d. (2004, October). Public displays of connection. *BT Technology Journal*, 22(4), 71–82.

Davies, J. (2004). Negotiating femininities on-line. *Gender and Education*, 16(1), 35–49.

Davies, J. (2006a). Affinities and beyond!! Developing ways of seeing in online spaces. *E-learning–Special Issue: Digital Interfaces*, 3(2) 217–234. Retrieved 24th December 2007, from http://www.wwwords.co.uk/elea/

Davies, J. (2006b). Escaping to the Borderlands: An exploration of the Internet as a cultural space for teenaged Wiccan girls. In K. Pahl & J. Rowsell (Eds.), *Travel notes from the new literacy studies: Instances of practice*. Clevedon Multilingual Matters. pp. 57–71. Davies, J. (2006c). "Hello newbie! **big welcome hugs** hope u like it here as much as i do!" An exploration of teenagers´ informal on-line learning. In D. Buckingham & R. Willett (Eds.), *Digital generations*. New York: Lawrence Erlbaum Associates. pp. 165–175. Davies, J. (2006d). Nomads and tribes: Online meaning-making and the development of new literacies. In J. Marsh & E. Millard (Eds.), *Popular literacies, childhood and schooling*. London: RoutledgeFalmer.

Davies, J. (2007). Display; identity and the everyday: Self-presentation through digital image sharing. *Discourse, Studies in the Cultural Politics of Education*, 28(4), 549–564.

Davies, J., & Merchant, G. (2007). Looking from the inside out: Academic blogging as new literacy. In C. Lankshear & M. Knobel (Eds.), *A new literacies sampler*. New York: Peter Lang. pp. 167–198.

Dawkins, R. (2006). *The selfish gene. 30ᵗʰ anniversary edition*. Oxford, UK: Oxford University Press.

Dowdall, C. (2006). Dissonance between the digitally created worlds of school and home. *Literacy*, 40(3), 153–163.

DrRob. (2005). *Poem for SimplyClare and drjoolz*. [Blogpost]. Retrieved 11th September 2006, from http://docrob.blogspot.com/2005/11/poem-for-simply-clare-and-dr-joolz.html" \l "comments

Facebook. [Social networking site]. Retrieved 26th July 2008 from http://www.facebook.com/

Fernback, J. (1997). The individual within the collective: Virtual ideology and the realization
of collective principles. In: S. G. Jones (Ed.), *Virtual culture: Identity and communication in
cybersociety* (pp. 36–54). London: SAGE.

Gee J. P. (1996). *Social Linguistics and Literacies: Ideology in Discourses* (2nd ed). London and New York: Routeldge/Falmer.

Kenway, J., & Bullen, J. (2001). *Consuming children: Education—entertainment—advertising*. Buckingham, UK: Open University Press.

Knobel, M., & Lankshear, C. (2007). *A new literacies sampler*. New York: Peter Lang.

Knobel M., & Lankshear, C. (2005, November 30). Memes and affinities: Cultural replication and literacy education. Paper presented to the annual NRC, Miami. Retrieved 11th September 2006 from: http://www.geocities.com/c.lankshear/memes2.pdf

Kress, G., & van Leeuwen, T. (1996). *Reading images: The grammar of visual design*. London: Routledge.

ljc@flickr. (2006). Frog Cupcakes. [Electronic image]. Retrieved November 15, 2007, from http://www.flickr.com/photos/ljc_pics/35664186/

Lankshear, C., & Knobel, M. (2006). *New literacies: Everyday practices and classroom learning* (2nd ed). London and New York: Open University Press and McGraw Hill.

Markham, A. (1998). *Life online: Researcjhing dxperience in virtual space*. London: Sage.

Mindscape. (1999). *Babyz*. [Computer software].

MySpace. [Social networking site]. Retrieved 3rd August 2006, from http://www.myspace.com/

Nielson/Net Ratings. (2007). *Young women now the most dominant group online*. [Online report]. Net Ratings UK Ltd. Retrieved August 30, 2007, from http://www.nielsen-netratings.com/pr/pr_070516_UK.pdf

O' Reilly, T. (2005). *What is Web 2.0? Design patterns and business models for the next generation of software*. [Blogpost]. Retrieved 11th September 2006 from http://www.oreillynet.com/pub/a/oreilly/tim/news/2005/09/30/what-is-web-20.html

Prensky, M. (2001). *Digital natives, digital immigrants*. Retrieved 24th December 2007, from http://www.marcprensky.com/writing/Prensky%20-%20Digital%20Natives,%20Digital%20Immigrants%20-%20Part1.pdf

Ruminatrix. (2006). *sockergubbarna befrielsefront*. [Set of electronic photographs]. Retrieved 24th December 2007, from http://www.flickr.com/photos/ruminatrix/sets/72057594127030623/./

SimplyClare. (2005). DrJools snapshotz on life. [Blogpost]. Retrieved 23rd December 2007, from http://drjoolzsnapshotz.blogspot.com/2005/08/just-to-add.html#112308222208016122

Street, B. (2003). What's "new" in new literacy studies? Critical approaches to literacy in theory and practice. *Current Issues on Comparative Education, Teachers College, Columbia University, 5*(2), 77–87.

Sutton-Smith, B. (1997). *The ambiguity of play*. Cambridge, MA and London, UK: Harvard University Press. p. 8.

Through the Viewfinder. (2006). [Set of electronic photographs]. Retrieved 24th December 2007, from http://www.flickr.com/groups/throughtheviewfinder/

TroisTetes. (2006.) The great escape. [Set of electronic photographs]. Retrieved 9th July 2007, from http://www.flickr.com/photos/trois-tetes/sets/72057594126052868/

Walker, J. (2004). *Distributed narratives: Telling stories across networks*. Paper presented at AoIR 5.0, Brighton, September 21, 2004. Retrieved 23rd March 2005, from http://jilltxt.net/txt/distributednarrative.html

Youtube. [Online videosharing website]. Retrieved 9th July 2006 from http://www.Youtube.com

8 Mimesis and the Spatial Economy of Children's Play Across Digital Divides
What Consequences for Creativity and Agency?

Beth Cross

> *The imagination makes unique models of the world, some of which*
> *lead us to anticipate useful changes, and some of which provide free-*
> *dom in their mockery of the constraints of the ordinary world....*
> *The flexibility of the imagination, of play, and of the playful is the*
> *ultimate guarantor of our survival.*
>
> (Sutton-Smith, 1997, p. 149)

INTRODUCTION: MIMICKING THE DIGITAL

This chapter draws on ethnographic and participatory research with children. The study was conducted intermittently over a three year period with
a group of 9–11-year-old boys who had an avid interest in the developing
media of computer based gaming. However, the focus of the chapter is on
what children do off screen. I follow what children borrow from digital
genres, and the ways they negotiate its use with each other in the play spaces
around their school and homes. In looking at how they mimic game genres
I draw on a dialogical understanding of mimesis within identity formation.
The chapter shows how children with different contexts for play borrow
from games in their play activities in different ways. I argue that digital
divides open up in different ways for children depending on the different
combinations of opportunities and resources available to them. Those children with more resources diverge from a strict or condensed imitation of
game features to a far greater degree than those with more constrained
times and spaces to play. The coincidence of the increased importance of
digital access at a time of shrinking opportunities for other kinds of playful
independent social interaction for children, it emerges, intensifies the disadvantages for children with spatially constrained resources. Through this
contextual study I explore what this means for how children experience
their own imaginations drawing on the work of Bakhtin (1981), Hughes
(1988) and Sutton-Smith (1997).

Hughes' (1988) suggestion that the imagination bridges internal and external reality is an important claim to consider and is the core concern of this study. It is this bridge between inner and outer, social and personal about which the least is known within game space, particularly as children are concerned. Children's agency, although of increasing focus, is still not understood in many respects. Buckingham (1999, 2000, 2003), one of the most cogent analysts of media and children, has drawn repeated attention to the lack of literature exactly at this interface. He argues that both those who celebrate and denigrate media influence on children do not pay enough attention to how media engagement is contextualised within the rest of a child's life. Prout (2005) raises very similar challenges and argues for an increased attention to a neglected middle ground between disciplinary concerns and approaches.

Although technological advancements have enabled a frame by frame observation of how children interact with the technology itself, measuring each key stroke and flicker of the eye, none of this tells us anything about what is going on in the interior imagination while its exterior counterpart is so absorbed. Studies of the aggressive consequences of viewing screen violence focus on immediate subconscious or unintended reactions, not on the intentional incorporation of its resources into play practices, or the relations there may be between the initial and the deliberate appropriation of its idioms. My interest therefore is to look off screen, to see what reinterpretation of gaming experience surfaces in children's other engagements with friends, objects and environments, an area with is underresearched (Holloway & Valentine, 2003, p. 124).

Over a three year period I observed a group of five boys in their middle childhood years from ages 9–11 as they incorporated material from screen play into their interactions across the communal play space that existed within the apartment complex where they lived. I reviewed these observations with the boys in a number of informal conversations. They also helped by drawing maps of some of their use of play spaces and explaining particular moves or strategies that were not easily read by the uninitiated. The chapter also draws from research undertaken in the playground of one of the schools the group of boys attended. Observations of play in the playground (of children aged 9–11) during recreational breaks was observed and these observations compared with those the playground attendants themselves offered when interviewed. In addition, interviews with staff in the after-school care club attached to the school are also drawn upon, as well as a focus group discussion with self-selected game experts among the children attending the club (children aged 9–12). The after-school club was attended by one of the boys in the longitudinal study which primarily served working class and lower middle class families, particularly those of single parent households. These observations were made in order to gain a sense of the broad range of play opportunities available more widely, particularly as the school served children with a broad range of socioeconomic backgrounds. It should be acknowledged, however, that the research engagement with children in these comparison sites was truncated. Deeper insights into the significance of their play repertoires would have been gained by further research.

Throughout research I strove to sustain what Christensen and Prout (2002) have called ethical symmetry, which means finding ways to respect children's interests and agendas, whilst exploring with them those of my adult standpoint. Sensitivity was required to find the fine borderline between inclusion and intrusion.

Based on observations of their use of these resources and on interviews with them, I was able to make a comparison to the arenas for social play that school playground and after school care facilities provided. For many children these more constrained environments constitute the only chance to share their versions and revisions of games with their peer group. The questions that guided this study were these:

- What are the images and scenarios from game space that children take up in off screen play?
- How are the meanings and dynamics of game space renegotiated through off screen play?
- How does different access to time and space impact on what children are able to express and explore?
- With what consequences for the relationships and identities children form through play?

THE MEANINGS OF MIMESIS IN CHILDREN'S PLAY

The crucial activity I was examining is imitation, or mimesis. A word here needs to be said about both the primacy and complexity of this activity. Hermans reminds us that imitation is the first sign of sociability we display in infancy, and describes it as 'the most rudimentary form of dialogical activity' (1999, p. 259). Here Hermans is alluding to the possibility that mimicking is not only a way of repeating what someone has done or said, but also of interpreting and commenting upon what is mimicked. In looking at mimesis in children's play repertoires Goldman (1998) draws on Coleridge's distinction between two kinds of imitation: catatropic, which strives to faithfully reproduce as exact a copy as possible; and metatropic, which inserts into the act of imitation deliberate meaningful deviations. These deviations insert the mimic's own perspective or intentionality into their reproduction. It is metatropic imitation that begins the dialogicality that within Vygotskian (1986) terms is the driver of a person's growing understanding of both themselves and those to which they relate. Metatropic imitation is the first skill from which many other metaskills are derived and remains a primary means of exploring and expressing, for instance, metacognitive and metalinguistic skills. Bakhtin's (1981) takes on language play also gives mimicry a primary role. He is emphatic that we are constantly borrowing our words, but again, rarely borrowing perfectly or fastidiously. To paraphrase him, always what we borrow we put to our own uses, say with our own inflection and fill with our own connotations in order to respond from our own specific purchase on events.

It is important to consider the power relations within mimesis, particularly as a form of dialogue. Metatropic mimicry by its very nature is a process open to variable interpretations, and therefore a useful strategy of those in subordinate positions. This may be one of the reasons it is such an attractive tactic for children. It works within the power structures without overtly challenging them. As a dialogic move it is not openly confrontational as it is never explicit. One's intention in mimicking can vary from: 'let me see if I get you right', 'let me see if it makes as much sense to be this way as your stance asserts', 'let me see if I can change it to suit myself without openly confronting you' to 'let me see if I can destabilise the power you hold in taking this position by demonstrating some of its weaknesses' (i.e., by parodying it, etc.). A child can slip between these various imitative intents and thereby play with the rules of relating without ever overtly challenging those with more power, or even explicitly admitting to self or other this is a possibility. As Sutton-Smith observes about the role of mimesis within play:

> Thus one can enact something real in the world, and thus be both innocent and guilty at the same time, only the shared knowledge of secrets allows others to know which truth if either is most intended. (1997, p. 139)

Mimesis is a way of sharing the world while also reserving for oneself an internal control over a version of it. As part of an identity integration process, it is a crucial staging ground.

Children's awareness of and participation in economic activity (Kenway & Bullen, 2001) is inherently caught up within digital game play and the imitation or borrowing of its resources off screen. In playing off screen children are not only consumers, but, using Hall's (1980) conception of articulation, are generators of new concepts and activities that precipitate further production–consumption cycles. Katz (2005) draws our attention to the significance of what may seem small inconsequential acts of mimesis in children's play repertoires in larger processes of reproduction, particularly the reproduction of class inequalities. Within play she sees the possibilities for resistance and resilience that reclaims agency as well as observing how constrained conditions reinscribe limitations on these. The relation between what is imaginatively possible and what is externally imposed is of crucial importance here.

NOT JUST ANOTHER WORLD, A DIFFERENT
GALAXY IN THE BACK GREEN

Let me then contrast the differing contexts in which I observed boys playing with resources they borrowed from screen engagement. The communal
play space that the apartment complex provided a small group of boys was
unique for an urban context.

Though I refer throughout to their space by the common Scottish usage
of back green, it was more like a park land with sloping hills, grand trees,
sheltered hollows and networks of labyrinthine passages beneath leafy
thickets. The complex was built in the walled grounds of an elegant estate.

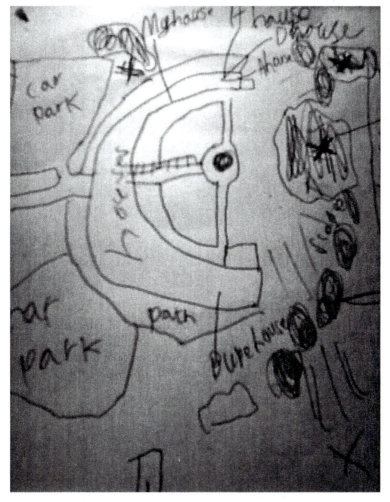

Figure 8.1 Play diagram of grounds surrounding apartments.

The landscaper of the complex had retained several of the features, including exotic plants from what would have been the British Empire when it was originally planted. Ironically, this same geographical diversity was represented in the population of the apartments' current tenants, all of them international postgraduate students and their families drawn from the University's contacts across Europe and the Commonwealth.

As this map illustrates, the housing consisted of a horseshoe of flats set in the middle of the ground. The primary play space consisted of a hill sectioned by walks leading to entrances to the stairways of the complex. This space was overlooked by the kitchens of all the flats, where parents could both work and keep an eye on the activity outside. However, the grounds were ample enough for there to be hidden play spaces under the trees, indicated by the scribbled circles that extended around the sides of the complex. The x's on one boy's representation of the play space indicate the different sites for staging the games they developed. The darker the x, the more frequent the use of the space.

Although the boys played a wide variety of available games for both game console and computer including games that focussed on physical skill and reflexes such as racing or shooting games, the boys expressed the most excitement about games that incorporated forms of strategic thinking that quest games such as *Zelda* required. It was these games primarily which served as the basis for their outdoor play. In this shared green space a few boys for hours at a time negotiated a wide variety of game scenarios. These games centred on duelling, often with sticks, sometimes with the dynamic powers of dragons or Pokémon-like pets. However, a great deal of time was spent in mutually narrating or dramatising quite complex story lines in which challenges and problems were devised and resolved. Thus moments of intense physical exertion including chases and battles were interspersed by long segments of leaning amongst the branches of trees exchanging in-character dialogue. In discussion the boys expressed a preference for games that involved pets that included elements of nurturing, training, marketing as well as competing, perhaps drawing on the format of *Sonic the Hedgehog*. These games afforded much more scope for creativity and a wider range of roles, and, it could be argued, provided a format for gaining a meta-awareness of their characters' social interactions, as the duelling and competing were once removed from the role they themselves adopted as farmers or owners in the games. Combat and pet games were played out over an extended period and included playing roles over a lifespan from the character's childhood and adolescence to adulthood. In a few cases they even took on the identity of old age characters who acted as advisors to the rest of the clan. These games played at life with all its vagaries including marriage, and parenting. The games usually centred on the development of a community between the characters and a threat or challenge to that community. They devised several different outcomes, including the complete destruction of their homes in one scenario and enforced flight to another land. The intricacies of the games, and the coordination they entailed are encapsulated in one boy's comparison of the differing dynamics within different game genres:

> M: Sometimes one person has to go in,
> and it spoils the whole game.
> Sometimes you play on
> and the other person doesn't know what has happened,
> doesn't always get the same stuff,
> so he gets a bit confused and the game starts to ruin.
> But with Dinosaur it's easier
> you can get other people to look after you pets
> while you go in.

Through the trial and error of these games, preferences for playing with violence were forged. Another boy commented:

> R: We don't like to play guns,
> it's so boring there is actually no fun in it–
> if you have a sword it's a challenge,
> they have a chance at you, it
> could go either way
> no matter what their training,
> but with a gun it's just at a distance and there's no chance.

The economy of their play engagement operated in tandem with that of game production. For weeks there would be very little use of outdoor space when an upgraded or novel game was being digested in feverish group engagement around the consoles of the new owners. Yet once a certain saturation point was reached, a period of exploration with new formats, weaponry, powers, and conflict would begin and the outdoor exploration would predominate, individual screen play being taken up as a second option only if others were not available. *Dynasty Warriors* was cited as particularly suited to their needs, trumping later releases. New game releases would generate an attraction back to the screen. How strong the pull depended upon a number of factors, to do with the innovativeness of the game, the ease of mastery, and the state of play in the back green at the point of interruption.

These games were rich examples of metatropic mimesis, involving extensive borrowing from game genres, but, as these remarks make clear, within a selective process of adapting features of the game that suited their tastes and needs. They drew not exclusively on games, but on toys (beyblades) and even book sources. One of the longest running games was based both on information gleaned from a nonfiction book on dinosaurs and on the groups' familiarity with Pokémon from an array of sources: cards, comics, videos, and hand held console games. The boys' reference to Pokémon indicated the genre was so pervasive that this constituted background knowledge rather than focussed intentional appropriation.

In speaking about their learning processes within play, the boys recognised a mimetic process at work:

> D: At first some of us didn't have a good imagination,
> but better than normal people. The others ones did have good
>   imaginations
> and they taught those that didn't
> Researcher: They copied off of R?
> D: Yeh, they got if off of R.
> Researcher: And then they made their own variations?
> D: Yeh.

One boy employed an imaginative reworking of a common toy genre in order to find terms to value the imaginary work that the back green play required: 'If R. was in top trumps he would be tops for imagination'. However the quote also highlights that digital resources alone did not ignite the imaginative interaction. If it needs reiterating, there were important social and cultural resources and skills that set this interaction apart from those widely practised among the local peer network in the more circumscribed contexts. All of the children had highly educated parents, with flexible working hours, who inevitably influenced the cultural capital that the boys drew upon in their play.

The boys' remarks on what they explicitly or implicitly incorporated into their role plays open an array of complex issues to assess. Primarily they spoke of what they borrow from games and game artefacts as content. The boys spoke of the imaginative work to create character and plot development with a greater sense of originality, as something they came up with themselves. When I asked if adult characters within the game treat children characters the same way that adults treat children in real life, they vigorously resisted the implication that they were borrowing from the social roles they are familiar with, and thereby internalising and reproducing them:

> D: You don't see,
> you have to imagine,
> this is nothing to do with humans,
> it's a different galaxy,
> not a different world, a different galaxy.

This response reflects the degree to which the boys saw their play scenarios as resisting the bounds of the domestic sphere within which they were enacted. It also indicates a sense of conflict between adult and child understandings of play or, to use Sutton-Smith's terms, a disagreement about the rhetorics of play that deserves further attention.

Throughout conversations with the boys who shared a back green, it emerges that this kind of imaginative play is not valued amongst their wider peer group, where it would be construed as babyish or unmasculine. However keen economic planners may be for Britain's up and coming workforce to embrace creative (i.e., entrepreneurial) identities, this is not a discourse the boys could access or use to legitimate to their larger peer group what they avidly valued in the protected sphere of the housing complex.

POKÉMON ON THE PLAYGROUND

My observations of digitally inspired game play on the playground revealed important differences to that developed in the back green. In the playground, boys engaged in short physical bursts of activity, making thrusting gestures that suggested duelling strategies. Often these ended in clenches and tussles that quickly broke apart. These exchanges were fairly rapid. The bouts followed each other in quick succession. Bouts were interspersed with brief exclamations before beginning again. These exchanges seemed to me to take about the same time as a digital game does to reset after a completed round of combat. This timing represented a fairly straightforward mimicking of the rhythm of fights within game formats.

Tight clusters of boys threaded across the playground, marking out through their physical activity the area they wished to preserve for their use. Sound effects and shouts predominated as interlocutors struggled to be heard amongst the competing activities of the crowded play space. The number of players was quite sizeable (15–20 players) depending upon the pull of other activities. Less commonly shared knowledge of games, and the constrained space meant that it was largely combative games played in close quarters, with easily readable gestures that could be sustained. However, playground monitors commented that role playing these action based games was an example of children working out roles and conventions, adapting them and arbitrating differences to a much greater degree than in other activities. As the adults were often not familiar with the rules or conventions of the digital games being adapted on the playground, they did not feel able to intercede as they could if there were disagreements about football or other traditional playground games.

The other opportunity the majority of boys had to socialise around game play was at the out of hours care club. The majority of the time children had there was shaped by the planned activities such as board games, craft activities, football or other team sports. The club had a policy of offering children different activity options. Although one of these options was playing console games, there was very little space for free play that might allow incorporating digital resources into off screen play, thus I did not have an opportunity to observe any off screen role play at the club.

Focus group work with a group of boys at the out of hours care facility revealed more about the constraints there are on describing the value of game play than it did about the boys' play experience. They spoke at length about what it was like to take turns at the consoles and the unevenness of time-keeping around turns. When I asked if there was any benefit to watching each other play, they allowed there was some, but this was counterbalanced by the frustration of watching weaker players make repeated mistakes. When the topic of learning was sounded out further, the boys associated this with educational games that covertly teach spelling or math skills. When asked about the games they prefer to play they asserted these do not have learning value. However, when I gave several instances of possible learning outcomes within *Grand Theft Auto* or *Zelda* (i.e., map reading, strategy formation, visual memory), they quickly affirmed those kind of things happen when they play, voicing at the same time a kind of surprise that that should count as learning. When asked if they learned things from playing the game off screen, again the boys found this odd to discuss. One boy saw playing off-screen in terms of how it could help him on-screen but concluded it wasn't often that something somebody did on the playground applied to on-screen. The primary benefit for playing off-screen parroted the prevalence of health messages now directed at children: 'on the playground you are using your energy and your friends.' Yet the impression remains that restricted, supervised activities constrained not only the interview, but the children's experiences discussed within it as well.

One instance of this restriction was in the kind of digital play provided for at the club. The club itself at this point in game development invested in games that focussed primarily on racing and sports. They were multiplayer in the sense the screen could be split into four which allowed for children to play simultaneously. However, these games facilitated very little interaction between players. What was really being facilitated was parallel play, not interactive play. At most a player could see another player race by, but there was no facility for interacting with each other. The more sophisticated, strategic digital play that did facilitate social interaction at that point had developed on single player handheld consoles such as gameboys. However, the club limited children's access to their hand held consoles to the last half hour of club out of a concern that children should socialise at the club and not isolate themselves by absorbing themselves in what appeared to be a solo activity.

This desire to help children socialise actually prevented them from engaging in a more sociable use of digital technology. It stems from reading spatial relations without factoring in the different dynamics of the digital dimension. I want to stress that this is very hard to do; it requires an imaginative leap that one can only make if one's own immersion within digital landscapes gives one the familiarity of experience to do so. Had the play workers understood children's hand held console play as social, it still leaves

Table 8.1 Spheres of Game Play

	Playground	Housing Complex
Number of Players	Variable number of several players	Small group of regular players
Quality of Space	Supervised constrained space, with numerable interfering factors.	Extensive terrain with little interference.
Characteristics of Enactment	Enactment remained close to easily inter-preted screen versions.	Enactment used screen resources adaptively and selectively and depended upon a 'game master' to generate detailed extensive scenarios.
Resulting Play Dynamics	Repetition of short combat scenarios drawn from fighting games.	Lengthy complex combination of duels, narration and life projects incorporating a larger mix of games that were predominantly strategy and quest oriented.

unresolved differences there may be between digitally mediated sociability and embodied sociability. Children's own bedrooms and homes provide them with access to digitally mediated sociability. Embodied sociability is becoming an increasingly scarce commodity. Due to the restrictions of children's supervised activities, scarcer still are opportunities for children to socialise both digitally and with peers simultaneously, the very opportunities they need to work out for themselves what differences, similarities and relative advantages and disadvantages both have.

To summarise, the main differences that emerge between children whose primary opportunity to play off screen are supervised play activities and those who supplemented this with extensive access to unsupervised play are outlined in Table 8.1.

THE IMAGINATIVE DIVIDE IN DIGITAL TERRAINS

The contexts discussed here suggest that a real appreciation of the kinds of digital divides, that is, the changing social demarcations that are further propelled by digital technology, requires us to look off screen, between screen and within concrete geographies, as well as across different uses of the technologies themselves. The children that made the most imaginative and socially enriched use of both their digital and concrete contexts were those who had a great deal of cultural and social capital. Moreover their highly educated parents encouraged them to engage in strategic games and activities with an eye towards play activity that would increase critical thinking skills and ameliorate the antisocial aspects of more rudimentary combat-oriented games.

The majority of children attending the school only had the condensed possibilities of playground space in which to work out social variations of screen play, as parents' long working hours and increased anxiety about the riskiness of public space meant children spent much more time in supervised activities or their own bedrooms than in any unsupervised self-directed communal peer space (Valentine, 2004). The back green play was a privileged setting in many respects. Access to any play space, let alone one as richly variegated in its landscape, was a rarity, and is becoming increasingly so. Working class single parents doing their best to conform to acceptable forms of childcare, with little time or resources to draw on to enrich their children's digital use are unlikely to be able to offer their children expanded play opportunities. If predictions of an hourglass economy are correct, and the steep curve between a large pool of service industry low paid jobs at the bottom of the economy and a smaller sphere at the top of highly paid, highly demanding jobs drawing a great deal of both creative and critical thinking skills becomes even steeper, this sector of society stands to lose the most in the deepening class divide. Although play may not seem immediately relevant, I would argue the differing experiences within this study are precisely the lived realities through which that divide is decided.

These concerns need to be qualified, as the interface between game space and play space will continue to change as the technology itself changes. The games industry is developing so quickly that children who began engaging with it ten years ago face a very different array of options and experiences than do those 5 years ago, or those entering it now. Earlier ethnographic research (Cross, 2005) showed boys losing interest in console gaming as they advanced through their upper primary years, preferring the pitch to the screen, and concentrating their digital engagement on communications networking across mobile phones and instant messaging platforms. However in the intervening years a much more sophisticated blend of genres and media increases the interactive capacity of games, and the game-like qualities within communication networks. One of the important factors affecting the interface between children's digital play and real life social interaction is the lag between children's relatively rapid uptake of the technology's changing capacity on the one hand and adults' rather retarded ability to accurately read and supervise these changes on the other.

THE IMPORTANCE OF THE INTERIOR–EXTERIOR IMAGINATIVE TERRAIN FOR POLICY AND PRACTITIONER INTERPRETATIONS OF PLAY

As the United Kingdoms's Labour Party thinktank Demos' recently released study *Their Space—Education for a Digital Generation* again reiterates, schools continue to lag behind, investing money, but not taking structural or cultural risks in order to open up interactive processes that digital capacities could enable (Green & Hannon, 2007). One playworker's remarks, made in an interview undertaken at the out of hours care club, may typify the crux of the problem: 'I don't feel kids shouldn't play, but there should be limits'. How to judge limits when all the dimensions are not visible is the difficult task that has many adults uneasy about the new landscapes children are traversing. The hesitation expressed here may be about finding it difficult to imagine what is happening within children's imaginative experience of digital terrains. Without experiencing gaming it is difficult to have the confidence that what they can judge by outward appearances is indicative of what is really going on in the interaction between screen and imagination that they cannot see. How adults understand limits is different from the limits children experience and there needs to be more of a vocabulary to mediate between the two. To return to the central concern of this chapter, there is a lack of common terminology to describe the imaginative resources and processes that build a bridge between inner and outer realities as socially negotiated projects.

In closing, I want to return to the importance of this internal–external dynamic of play as a component of social processes. There are some important correlations between Hughes (1988) and Bakhtin's (1981, 1984) understanding of the relations between imagination, mimesis and social processes and Sutton-Smith's (1997) explication of the many different meanings and uses of play. An overview of them is provided here that it is hoped will provoke the readers' further exploration.

In his essay on myth and imagination Ted Hughes (1988) uses story to explicate the workings of the imagination. He argues that stories take on a life of their own once they have been internalised. Stories both expand into internal worlds but at the same time can be condensed and, like a seed, transplant themselves into the world of other stories. The most vivid events can be conveyed as a single image encapsulated in a single word which the imagination recombines with other such potent carriers of meaning, giving rise to hybrid confrontations and catalytic creations. This process relies fundamentally on metatropic mimesis and interpretation. In this model these transpositions are one of the primary activities of the imagination. Hughes continues in his essay to compare the externally focussed imagination to the imagination of the internal, voicing a growing concern that with the decline of religious myth we have lost a bridge between the two, and that both the imbalance and attenuation of relations between the two imperil both human sanity and society.

Bakhtin (1984) as well looks to the interrelation between what communities enact together and what is internalised from this social stage to an interior one. In his work on Rabelais he argues that carnival, with its exaggerations, caricatures, transpositions and, most importantly, role reversals, is integral to the wholeness and well being of both external and internal human relations. Metatropic mimesis is the primary vehicle through which carnival is accomplished. Critique of the powers that be and thereby an enlivening of relations between ruled and ruling is accomplished through parodies of the powerful and the enactment of their roles in such a way as to break the rules. Self and other and how we constitute each other is Bakhtin's primary aim of both this work on carnival and his examination of the dialogic possibilities emergent in the novel.

A consideration of Bakhtin's carnivalesque is incorporated by Sutton-Smith (1997, pp. 140–142) into his analysis of the different meanings play can have which he categorises as constituting different rhetorics of play. He points out the differing roles of mimesis within different rhetorics of play by observing that some play consists in playing within the rules and some play plays with the rules themselves. He aligns carnivalesque with the rhetoric of play as identity, and play as phantasmagoric. Both these forms of play involve the most extensive use of metatropic mimesis.

Highly critical of functionalist rhetorics of play, particularly those that subsume play into an education agenda, Sutton-Smith (1997) nevertheless in the conclusion to the *Ambiguity of Play* articulates an argument for the importance of play that responds to functionalist concerns. Play may be a safe place to do the experimenting that the species needs if it is to have the flexibility and adaptability to respond to the changing demands of our context, even if it is our own technological and socioeconomic impositions that have changed it. Playing with violence may be a mechanism for creating a wider variety of many possible scenarios, a fractal architecture of imaginative resource. The richer, the more transposed and original the combinations, the more flexible and adaptive real life responses may be. If one watches many of the Jackie Chan movies, in which he is always thrown into situations in which weapon less he must defend himself, one will recognise the ingenious use of objects in his environment that gives these movies their appeal and make them such a rich source for mimetic reworking by children.

Sutton-Smith (1997) distinguishes between children's storytelling and play, but like Hughes, sees story as an important process that sheds light on the playful workings of the imagination. In looking at the cataclysmic bent of children's early story making Sutton-Smith observes, 'it may be pertinent to think of all early stories and all early play as cyclical forms of virtual transcendancy' (1997, p. 164), the point being not to arrive or achieve but to perpetually resuscitate a state of absorption, to stay in the midst of the journey, or, I propose, to remain in a space that explores the interface between internal and external reality. Mackey (this volume) makes a similar argument for the appeal of the 'unfinished'.

Sutton-Smith relates that by the time children are 11 they are producing stories in which a hero has emerged to save the day with miraculous happy endings. In previous fieldwork (Cross, 2005), I observed that the further the boys I worked with felt they could stray from the expected norms of the curriculum, the more they dispensed with such niceties with gusto, suggesting that this adoption of a linear progressivist form may be more of a concession than an internally motivated development. Children's practices in that study as well as this one strongly evidenced that children's strategies for borrowing and rejuxtaposing cultural resources exceed the normative frames used to value their creative merit or educational worth. These finding corroborate the critique Sutton-Smith (1997) repeatedly makes in the *Ambiguity of Play* of the pervasive educational rhetoric that appropriates play for rationalist development goals at the expense of its more illusive qualities. In sanitising play we may be expunging healthy growth at a level we are as yet ill-equipped to measure or judge.

I concluded from my examination of the communal play with game and story examined here that they weren't trying to make meaning, to condense possibility down to a workable, useful survival strategy. That is only one of the uses of narrative and one focussed on to the exclusion of the potential Hughes and Bakhtin see within it. It can also be a tool of exploration and regeneration, in which the object is not to decide how reality is but to explore it. Sutton-Smith posits a parallel play world in which play is not used to rehearse or recover from real life trauma, as it is theorised in psychotherapy, the one area of psychology to pay attention to the dynamics of children's play worlds in any detail, but as phantasmagoric indulgences for their own sake. The indulgence in this excessive violence may not be to rehearse fear so as to learn how to survive it. Bolstering the belief in our survivability, handy as that is, may be nothing more than a useful side effect, but not the impetus or internal engine which drives it.

As Sutton-Smith points out, trying out all sorts of transpositions and rearrangements in real life would be highly dangerous, costly and disastrous, so there may be good reason why it is relegated to an activity about which we are dismissive, the play of children. This may be one reason why there is such a disjuncture between play with its suspended cyclical quality and purposeful activity which enjoys no such suspension. Attempts to harness play to the needs of the market or education may be threatening our ability to sustain an internal or imaginative biodiversity far more than we realise. Increasing trends which marginalise opportunities and spaces for unsupervised social play seem likely to contribute to widening inequalities. Spaces where children can access each other in real embodied time as well as digital media are least available, and yet offer children the greatest potential to integrate their play experiences. The digital divide is about inequality of access to a number of interlinked resources. This chapter has sought to explore the imaginative ones. Over the course of the ESRC seminar series, 'Play, Creativity and Digital Cultures', on which this book is based, several conversations have sought to make connections between the rhetorics that surround each of these key terms. Drawing on Sutton-Smith's categorisation and alluding to the work Banaji and Burn (2006) have done on the rhetorics of creativity, I suggest that the following linkages exist.

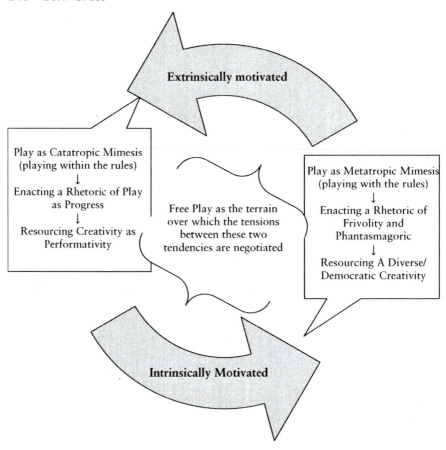

Figure 8.2   The relations between mimesis and the rhetorics of play and creativity.

As the diagram depicts, different rhetorics tend to align towards different outcomes. As the boys play with differing forms of battling and cooperating they are also negotiating a battlefield of rhetorics, with very real consequences for the kinds of identities and agency they will have recourse to throughout their lives.

In a sense, I am making a similar appeal for a public space for play that Arendt (1958) makes for a public space for reason. These spaces are not antithetical to one another. If my reading of Bakhtin and Hughes is correct, the health of both may depend upon each other more than we have allowed. This examination of terrain between digital and concrete play spaces in the end reiterates the importance of attention to children's spaces advocated by Moss and Petrie (2002). They argue that a concerted effort across professional specialities is needed to change the relations that are necessary to resource such spaces. This is not an educational question alone. In fact urban planning is as pertinent, if not more so, to the dynamics examined here. If it is true that 'We shape our buildings, and afterwards our buildings shape us' (Churchill, 1943), more attention needs to be given to how we provide for the imaginative space children need to contextualize their digital experiences. I agree with Sutton-Smith that this would require not just surface or minor procedural changes, but a change in the fundamental concepts that underpin policy making assumptions. Having begun with a quote from Sutton-Smith I close with another:

> Our children deserve an adult rhetoric that will pay respect to their use of play for both power purposes and phantasmagorical purposes. At the moment, the rhetoric of progress blots out the possibilities for such a larger child-oriented humanism. (1997, p. 166)

REFERENCES

Arendt, H. (1958). *The human condition.* Chicago: The University of Chicago Press.
Bakhtin, M. M. (1981). *The dialogic imagination: Four essays* (M. Holquist, Ed.). Austin, TX: University of Texas Press.
Bakhtin, M. M. (1984). *Rabelais and his world.* Bloomington, IA: Indiana University Press.
Banaji, S., & Burn, A. (2006). *The rhetorics of creativity: A review of the literature.* London: Centre for the Study of Children, Youth and Media.
Buckingham, D. (1999). Superhighway or road to nowhere? Children's relationships with digital technology. *English in Education, 33*(1), 3–12.
Buckingham, D. (2000). *After the death of childhood: Growing up in the age of electronic media.* Cambridge, UK: Polity Press.
Buckingham, D. (2003). Gotta catch 'em all: Structure, agency and pedagogy in children's media culture. *Media Culture and Society, 25*(3), 379–398.
Christensen, C., & Prout, A. (2002). Working with ethical symmetry in social research with children., *Childhood, 9*(4), 477–497.
Churchill (1943) House of Commons (meeting in the House of Commons), 28 October 1943.
Cross, B. (2005). Split frame thinking and multiple scenario awareness: How boys' game expertise reshapes possible structures of sense in a digital world. *Discourse Studies in the Cultural Politics of Education, 26*(3), 333–354.
Green, H., & Hannon, C. (2007). *Their space—Education for a digital generation.* Retrieved October 10, 2007, from <http://www.demos.co.uk/publications/theirspace>
Goldman, L. R. (1998). *Child's play: Myth, mimesis and make-believe.* London: Berg.
Hall, S. (1980). Encoding and Decoding, In Stuart Hall, Dorothy Hobson, Andrew Lowe and Paul Willis (Eds.), *Culture, media, language.* London: Hutchison, pp. 128–138.
Hermans, H. (1999). Dialogical thinking and self innovation. *Culture and Psychology, 5*(1), 67–87
Holloway, S., & Valentine, G. (2003). *Cyberkids: Children in the information age.* London: RoutledgeFalmer.

Hughes, T. (1988). Myth and education. In K. Egan & D. Nadaner (Eds.), *Imagination and education* (pp. 30–44). Milton Keynes, UK: Open University Press.

Katz, C. (2005). *Growing up global: Economic restructuring and children's everyday lives.* London: University of Minnesota Press.

Kenway, J., & Bullen, E. (2002). *Consuming children: Education, entertainment, advertising.* Buckingham, UK: Open University Press.

Moss, P., & Petrie, P. (2002). *From children's services to children's spaces: Public policy, children and childhood.* London: RoutledgeFalmer.

Prout, A. (2005). *The future of childhood.* London: RoutledgeFalmer.

Sutton-Smith, B. (1997). *The ambiguity of play.* London: Harvard University Press.

Valentine, G. (2004). *Public space and the culture of childhood.* London, UK: Ashgate.

Vygotsky, L. (1986). *Thought and language.* Boston, MA: MIT Press.

Part III

Play, Creativity and Digital Learning

Introduction to Part III

This section of the book starts with a close examination of notions of creativity and play, analysing specific constructions of these complex concepts; and then the section extends the discussions of play, creativity and digital cultures developed in the previous chapters to look specifically at questions related to learning and changing notions of literacy. In the first chapter in this section, Shakuntala Banaji suggests that even a cursory perusal of the history of research into creativity reveals several robust and persistent trends, a plethora of 'descriptive anecdotes' or reference to 'psychological characteristics' or 'processes'. Depending on the tradition to which the researcher belongs, these trends are associated with a series of political and philosophical presuppositions about human beings and society that are seldom traced back to their historical roots. Recent trends see creative activity as both a cure for the ills of an increasingly troubled society, and as a charm to unlock the potential and boost the morale of demotivated and excluded sections of children and youth, the populace, the community or the workforce. Via an exploration of a number of contemporary and persistent political and philosophical traditions in the theorising of creativity, this chapter asks: To what extent are any of these claims a reflection of actual events, trends and practices? Whose interests do some of these conceptualisations serve? And is there any point in trying to find a singular definition that encompasses the many 'creativities' that have been conceptualised?

In Chapter 10, Caroline Pelletier examines the forms of meaning which a group of nine young people aged 12 produced in planning the design of their own video game in an after-school club. Planning work involved playing, analyzing and editing board games and considering broad principles of game design which apply across platforms. Describing an action or thought as a process of play locates it within a particular interpretative frame; it indicates what that action or thought means, and situates it in relation to other kinds of (nonplay) activities. One might therefore define play as a particular way of creating meaning; it enables distinctive forms of communication and representation, or allows particular things to be perceived, said and understood. Caroline focuses on conceptions of play, and more particularly the forms of representation and communication that play made possible. This chapter highlights the social function that play performs in negotiating social relations.

In Chapter 11, Victoria Carrington uses examples drawn from a range of digital technologies to explore two interesting social phenomena: the emergence of new forms of civic participation and shifts in our view of play and creativity. An exploration of these two currents informs an analysis of the influence of digital technologies on the everyday life of our society and as a consequence, their significance for school-based literacy practices.

In the final chapter of the book, Jackie Marsh develops this theme by focusing on the way in which classroom pedagogies need to evolve in order to ensure that schooling is relevant to a generation of learners who are growing up in the digital era. Engaged in a wide range of playful and creative digital practices outside of school, pupils need to engage in challenging and productive activities in schools in order to ensure effective and meaningful learning. Drawing from a project in which a primary teacher introduced children to the practice of blogging, the chapter demonstrates how, through the facilitation of playful approaches to a digital literacy curriculum, educators can develop a pedagogical approach that ensures learner agency, curriculum relevance and intellectual challenge.

9 Creativity
Exploring the Rhetorics and the Realities

Shakuntala Banaji

INTRODUCTION: THE RHETORICS OF CREATIVITY[1]

This chapter explores understandings of creativity in relation to social relations, play and pedagogy in policy and practice: where these understandings come from in terms of their theoretical heritage, what functions they serve, how they are used, and in whose interest. The focus is on discourses about creativity circulating in the public domain. The aim here is not to investigate creativity itself, but rather what is written and said about it. Creativity is thus presented here as something *constructed through discourse*, and the ensuing discussion aims to envision more clearly how such constructions work, what claims are being made, and how we might choose to locate ourselves in relation to these claims. In the critical review of literature from which this chapter originates, (Banaji & Burn 2007b), the rhetorics of creativity are given names which broadly correspond to the main theoretical underpinnings or the ideological beliefs of those who deploy them. Thus, the rhetorics referred to in this chapter are as follows:

- Creative Genius;
- Democratic Creativity and Cultural Re/Production;
- Ubiquitous Creativity;
- Creativity for Social Good;
- Creativity as Economic Imperative;
- Play and Creativity;
- Creativity and Cognition;
- The Creative Affordances of Technology; and
- The Creative Classroom
- Creative Arts and Political Challenge.

The rhetorics identified have complex histories, particularly in traditions of philosophical thought about creativity since the European Enlightenment and in parallel forms of artistic practice, and in traditions of educational theory and practice related to creativity and play over the same period. In coming sections, following brief historical descriptions, the rhetorics are traced through in academic and policy discourses or, via reference to other research, in the discourse of students and teachers.

The discussion of individual rhetorics raises a series of questions that cut across and connect several rhetorics to each other. For instance, two questions running through the rhetorics of Genius, Democratic and Ubiquitous creativity are: Does creativity reside in everyday aspects of human life or is it something special? And what are the differences between 'cultural learning' and 'creative learning'? Writing on creativity in education distinguishes between creative teaching and creative learning but often fails to establish precisely how such processes and the practices they entail differ from 'good' or 'effective' teaching and 'engaged' or 'enthusiastic' learning. Thus the issue of whether there is, in fact, any difference between 'good' pedagogy and 'creative' pedagogy is the focus of attention in a number of the rhetorics from those that see Creativity as a Social Good to those that deal with students and the classroom. Meanwhile, the questions of how significant play and individual socialisation are as components of creativity link rhetorics as diverse as those concerned with Technology and the Economy to Cognitive and Play theories.

More specifically, in the context of this volume, this analysis contextualises positions taken up by diverse groups of parents, academics, educators and policymakers with regard to the uses of digital technologies by children and young people. The concerns of those who view media technologies as inhibiting children's apparently 'natural' creativity and those who view poor teaching as inhibiting the creative potentials of technology are aired alongside accounts of the actual uses to which teachers and children are putting new digital technologies. The following sections lay the groundwork for this discussion by examining rhetorics about the nature of creativity, its potential transmission and measurement, as well as about the space for creative participation in varied economic and educational contexts.

CREATIVITY: UNIQUE OR DEMOCRATIC?

The rhetoric which could be said to have the oldest provenance and to have remained resilient, albeit in more subtle guises, within educational pedagogies in the 20th and 21st century, is that of *Creative Genius*. This is a romantic and post-romantic rhetoric that dismisses modernity and popular culture as vulgar, and argues for creativity as a special quality of a few highly educated and disciplined individuals (who possess genius) and of a few cultural products. Culture in this rhetoric is defined by a particular discourse about aesthetic judgment and value, manners, civilization and the attempt to establish literary, artistic and musical canons. It can be traced back through certain phases of the Romantic period to aspects of European Enlightenment thought. Perhaps the most influential Enlightenment definition of genius is in Kant's *Critique of Judgment*, which presents genius as the 'mental aptitude' necessary for the production of fine art, a capacity characterised by originality, and opposed to imitation.

Some contemporary commentators remain implicitly attached to the idea of genius (Simonton, 1999; Scruton, 2000). This view is interestingly at odds with a common definition of creativity as needing to involve novelty. In an essay entitled 'After Modernism' (2000), Roger Scruton draws a distinction between inspired and vulgar architecture:

> Our best bet in architecture is that the artistic geniuses should invest their energy . . . in patterns that can be reproduced at will by the rest of us. . . . In making innovation and experiment into the norm, while waging war against ornament, detail, and the old vernaculars, modernism led to a spectacular loss of knowledge among ordinary builders and to a pretension to originality in a sphere where originality, except in the rare hands of genius, is a serious threat to the surrounding order. (Scruton, 2000)

Notably, while the language used here counterposes the ordinary with the exceptional, there is a sense in which 'novelty' is viewed as a negative, almost dangerous, attribute when proposed by those who do not possess the requisite skill and inspiration to maintain a link with what is seen to be the best in the past. Scruton is not alone in his concerns about the debasement of 'real' art, the rejection of 'training', 'rules', 'traditions' and so on. Websites such as 'The Illinois Loop' (a supposedly critical look at school education in that state) pride themselves on taking issue with 'creative' aspects of the modern arts curriculum.

> When your 6th grader comes home and proudly shows you the "art project" he made in school from shoeboxes, duct tape, and spray paint, a valid question is, "Is my child learning anything about art?" In the context of the art program itself, the overwhelming emphasis in most schools is on art as a *hobby and craft*, with heavy favoritism of "creative" projects (painting an album cover, decorating a hub cap, etc.). *Yes, it's fun. And some of the projects are indeed delightful. And no one doubts that kids should have time to be kids and let their creativity thrive.* But what is missing? (http://www.illinoisloop.org/artmusic. html, emphasis in original)

The view of art as being about self-expression is derided as a mere loss of skill and in some cases as an apology for absent skills. Significantly for the rhetorics *Play and Creativity*, and *The Creative Classroom*, such discussions caricature the supposed 'opposition' and mobilise parental concern around a constructed binary opposition between 'pointless playing around' (creativity) and 'real learning' (academic progression within a sanctioned tradition).

Many educators and parents still operate within frameworks such as those outlined above. For instance, fears abound that allowing teachers and children time within English lessons to use or 'play' with mother-tongue languages, television programmes, blogs or to make other forms of digital media will mean a loss of all structure, traditional literacy and discipline. In many contemporary national educational contexts (Cremin, Comber & Wolf, 2007) policy reactions have tended to be against this caricature of 'everything goes' *laisse faire* rather than in the light of real classroom practices. It has been argued (Maisuria, 2005; Kwek, Albright & Kramer-Dahl, 2007) that recent trends in assessment in the United Kingdom and elsewhere tend to focus on the transmission and acquisition of isolated skills and bits of canonical knowledge, thus missing the long-term impact of creative learning experiences which can be assessed formatively through self-critique and joint commentaries.

In this context, attempting to make creative teaching more palatable to those who believe in canonical knowledge and a transmission-orientated curriculum, some commentators write as if there are two different 'categories' of creativity. These have been dubbed, variously, 'high' and 'common' (Cropley, 2001) or 'Historical' and 'Psychological' (Boden, 1990) or 'special' and 'everyday'. The former comprises the work and powers of those who are considered 'geniuses' in the rhetoric just examined. It is pursued via studies of the work and lives of 'great' creative individuals (Csikszentmihalyi, 1997) and is seen as being 'absolute'. The latter more relative notion—which argues that creativity can be fostered, increased and measured—can also, broadly, be split into two traditions: one grounded in culture or subculture; the other based on notions of 'possibility thinking' and dubbed little 'c' creativity (Craft, 1999).

Providing an explicitly antielitist conceptualisation of creativity as inherent in the everyday cultural and symbolic practices of all human beings, is a rhetoric relating to *Democratic Creativity and Cultural Re/Production*. This rhetoric, most familiar in the academic discipline of cultural studies, sees everyday cultural practices in relation to the cultural politics of identity construction. It focuses particularly on the meanings made from and with popular cultural products. This rhetoric provides a theory derived from the Gramscian perspective on youth subcultures developed by the Birmingham Centre for Contemporary Cultural Studies. It constitutes practices of cultural consumption (especially of films, magazines, fashion and popular music) as forms of production through activities such as music sampling, subcultural clothing and fan activity (Cunningham, 1998); and thus belongs to an influential strand of cultural studies which attributes considerable creative agency to those social groups traditionally perceived as audiences and consumers or even as excluded from creative work by virtue of their social status (Willis, 1990). In a different incarnation, it can be seen at work in the arguments of David Gauntlett (2007, pp. 19–25) who locates visual 'making' and communication (video diaries, building block constructions) as central to processes of creative identity expression.

Similarly egalitarian, but without the basis in cultural politics, is the rhetoric of *Ubiquitous Creativity*. Here, creativity is not just about consumption and production of artistic products, whether popular or elite, but involves a skill in terms of responding to the demands of everyday life. To be more creative, in this discourse, involves having the flexibility to respond to problems and changes in the modern world and one's personal life (Craft, 1999, 2003). While much of the writing in this rhetoric is targeted at early years education with the aim of giving young children the ability to deal reflexively and ethically with problems encountered during learning and family life, examples used to illustrate 'everyday creativity' include attempts by working-class individuals or immigrants to find jobs against the odds without becoming discouraged. This too is a resilient strand in commentaries on this subject and has a strong appeal for educators (Jeffrey, 2005; Cohen, 2000) who wish to emphasise the significance of ethical, life-based education that does not rely on the transmission of particular traditionally judged knowledge and skills.

Clearly for those who see creativity as something 'special' or 'arts-based', or indeed who see it as being about challenge and social critique, this approach remains problematic. Negus and Pickering (2004) develop a strong critique of little 'c' creativity, arguing that:

> . . . we cannot collapse creativity into everyday life, as if they are indistinguishable. . . . Only certain of our everyday experiences involve creativity; only some of our everyday actions are creative. . . . What we're arguing for instead are the intrinsic connections *between* creative practice and everyday life, for it's important that we don't forget how the heightened moments of creativity are always linked to routine and the daily round, and how a particular artwork or cultural product may catch us within the midst of ordinary habitual life. (2004, pp. 44–45)

While this view delinks creativity from mundane activities while allowing for its location alongside the everyday, it leaves in place the tensions between activities, ideas and products that are socially accepted as 'creative' in particular historical moments and those that are rejected for fear of their playful, disruptive or anarchic potential. Thus even the work of artists such as William Blake or political philosophers such as Karl Marx, while acknowledged by some as extraordinarily creative, has also been feared by many for encouraging uncharted, troublesome and subversive ways of feeling, thinking and relating within society. The following section explores further the implicit tension between a wish to foster the socially acceptable, benevolent effects of creativity (embodied in academic and economic success) and the current aversion in schools and academic institutions to pedagogic activity which encourages fantasy play or sociopolitical critique.

CREATIVE SOCIALISATION AND 'SUCCESSFUL' SOCIETIES?

The rhetoric of *Creativity for Social Good* sees individual creativity as linked to social structures. This rhetoric is characterised by its emphasis on the importance for educational policy of the arts as tools for personal empowerment and ultimately for social regeneration (the NACCCE report: Robinson et al., 1999). It stresses the integration of communities and individuals who have become 'socially excluded' (e.g., by virtue of race, location or poverty) and generally invokes educational and, tangentially, economic concerns as the basis for generating policy interest in creativity. This rhetoric emerges largely from contemporary social democratic discourses of inclusion and multiculturalism. In this view, a further rationale for encouraging creativity in education focuses on the social and personal development of young people. This encompasses a bow to multiculturalism (Robinson et al., 1999, pp. 22–23) and antiracism, as well as an avowed desire to combat growing drug use, teenage alcoholism and other social problems. In this view, 'creative and cultural programmes' are seen to be two-fold mechanisms of social cohesion, 'powerful ways of revitalising the sense of community in a school and engaging the whole school with the wider community' (Robinson et al., 1999, p. 26).

Although Robinson's NACCCE committee team accept that exceptionally gifted creative individuals do exist, their report favours a 'democratic' definition of creativity over an 'elite' one: 'Imaginative activity fashioned so as to produce outcomes that are both original and of value' (1999, p. 29). For them imaginative activity entails a process of generating something original, whether this be an idea or a relationship between existing ideas. This immediately sets it apart from discourses which might be seen to encourage a view of creative and imaginative activity as play or fantasy. The NACCCE report is implicitly suggesting that the preparatory and exploratory time in art, media, technology and drama classrooms and projects is only valuable insofar as it contributes to the final product or to the reinsertion of 'excluded youth' into the official school system. Culture and other cultures are things to be 'dealt with' and 'understood'. While this reductive view has been implicitly critiqued on various occasions (Marshall, 2001; Buckingham & Jones, 2001) it has a broad appeal amongst those who see creativity as a tool in the project of engineering a strong national society.

In an allied rhetoric, *Creativity as Economic Imperative*, the future of a competitive national economy is seen to depend on the knowledge, flexibility, personal responsibility and problem solving skills of workers and their managers (cf. Scholtz & Livingstone, 2005). These are, apparently, fostered and encouraged by creative methods in business, education and industry (Seltzer & Bentley, 1999). There is a particular focus here on the contribution of the 'creative industries', broadly defined, although the argument is often applied to the commercial world more generally. This rhetoric annexes the concept of creativity in the service of a neo-liberal economic programme and discourse (Landry, 2000). Indeed, although they claim to be interested in a diversity of contexts, flexibility of learning, self-evaluation and student empowerment, much of Seltzer's and Bentley's emphasis is directed towards getting more IT literacy and knowledge of computers into the curriculum and getting young people into industrial/business placements at an early stage, whether in school or university. Instead of being about imagination (which at least plays a role in the NACCCE report) or about the motivation to learn and create, the imperative here is the requirement to assist the modern national capitalist economy in its quest for global expansion.

Training courses in 'creativity', promising anything from personal fulfilment and office bonding to higher profits and guaranteed jobs, abound both online and offline.[1] But, realistically, we must ask questions about the variety of arenas and domains in which those who buy into this 'new' vision of creativity would be allowed to function. It is unlikely that time for playful testing of ideas would be built into the working days of 'knowledge workers'. Perhaps they would have to accommodate such necessary but peripheral business in their own personal time by giving up leisure (as is increasingly the case with the penetration of work-related ICT in the home). And in what way might different skills lead to creative production? It seems unlikely that the mere acquisition of skills would be enough as a contribution to a greater collective or corporate endeavour. Clearly, while the newly flexible workforce described by Seltzer and Bentley (1999) might be encouraged to manage themselves and their departments or sections, their control over the overall structures and practices of their organisations might remain as limited as ever. Indeed, as Rob Pope (2005, p. 28) poignantly describes with regard to two of the companies presented as shining examples of such newly creative practices in *The Creative Age*, jobs and livelihoods may be no more secure if workers become 'creative' and 'flexible' than those in very 'old fashioned' manufacturing jobs that did not fall within the scope of the so-called 'knowledge economy'.

SERIOUS OR PLAYFUL STUFF?

The rhetoric of *Creativity and Cognition* can be seen as incorporating two quite different traditions. One tradition includes theories of multiple intelligences (Gardner, 1993) and the development of models to document and increase people's problem solving capacity (e.g., Osborn-Parnes' 1941 CPS model) as well as explorations of the potential of artificial intelligence (Boden, 1990). This latter work attempts to demonstrate the links made during, and the conditions for, creative thought and production. The emphasis of all strands in this tradition is nevertheless on the *internal* production of creativity by the mind, rather than on external contexts and cultures. The other tradition consists of more intracognitive and culturally situated notions of creative learning expounded by Vygotsky (1931/1994), who asserts that: 'If a person "cannot do something that is not directly motivated by an actual situation" then they are neither free nor using imagination or creativity' (1931/1994, p. 267). He writes:

> Our ability to do something meaningless, absolutely useless and which is elicited neither by external nor internal circumstances, has usually been regarded as the clearest indication of the wilfulness of resolve and freedom of action which is being performed . . . [thus] . . . imagination and creativity are linked to a free reworking of various elements of experience, freely combined, and which, as a precondition, without fail, require the level of inner freedom of thought, action and cognising which only he who has mastered thinking in concepts can achieve. (1931/1994, p. 269)

The importance given to 'freedom' of thought and action and to non-goal orientated playful activity in Vygotsky's writing about adolescent learning remains controversial in educational or economic environments where the ability to plan a project and execute it, solve a problem or pass a test are markers of effectiveness. Controversially for some, the emphasis in this theorizing is on developing patterns of thought and conceptual understanding in particular social and cultural settings ('various elements of experience') rather than on mastering a canon or a body of knowledge.

What may be termed 'inter-cognitive' theories of cognition, spanning a spectrum from psychometric tests and scales to 'experimental' studies on groups of young people or lecturers, have been heavily utilized to 'prove' the existence and/or the level of benefit of retaining a place for 'creativity' in formal educational settings. More flexible indicators of creativity such as the various 'Intelligences' described by Gardner have been used on occasion in a positive manner to soften the harshness of traditional literacy and numeracy-based academic assessment. Vygotsky's far more critical and unusual theorizing, however, has been largely ignored by contemporary policy makers. Yet Vygotsky never denies that creativity has concrete results, or that these cannot at some point be evaluated. In fact, to him: 'It is the creative character of concrete expression and the construction of a new image which exemplify fantasy. Its culminating point is the achievement of a concrete form, but this form can only be attained with the help of abstraction' (1931/1994, p. 283). The point here is that creativity requires patience and an appreciation of the playful, and perhaps the fanciful and insubstantial.

A persistent strand in writing about creativity and intersecting another set of rhetorics that centre on childhood, the rhetoric of *Play and Creativity* turns on the notion that childhood play models, and perhaps scaffolds, adult problem solving and creative thought. It explores the functions of play in relation to both creative production and cultural consumption. Some cognitivist approaches to play do share the emphasis of the *Creative*

Classroom rhetoric on the importance of divergent thinking. Sandra Russ, for instance, argues that '[p]lay has been found to facilitate insight ability and divergent thinking' (2003, p. 291), and that 'theoretically play fosters the development of cognitive and affective processes that are important in the creative act' (2003, p. 291). Challenging a mainstay of the economistic conceptualisation of creativity, she sees children as being excluded by definitions of creative products as effective, novel and valuable. Like Carruthers (2002) she argues that the ways in which children use language, toys, role-plays and objects to represent different things in play are habitual ways of practising divergent thinking skills. Accounts such as these raise questions for those interested in creativity, pedagogy and learning. For example, there is widespread concern (cf. Brennan, 2005; Maisuria, 2005) about the way in which childhood pretence and play are being squeezed out of the school curriculum to be replaced by the learning of rules and appropriate roles, rote literacy tasks and an approximation of 'adult'-type problem solving tasks. Brennan suggests that 'pretend play . . . is both a learning and teaching tool' and asserts that 'a play curriculum inherently recognises the inseparability of emotion and cognition, and consequently of care and education and values the bioecological context in which both are embedded' (2004, pp. 307–308). The stories in her report which 'assess' children's learning and creativity through play are all told with a sensitivity to the contexts and relationships in the lives of participants, as well as the immediate tools and goals achieved. Most significantly, extrapolating from the manner of assessment described here, there is a recognition that playful learning and creative experiences form a continuum rather than being isolated units which can be measured and enumerated with any degree of authenticity. In such rhetorical constructions, opportunities for and contexts of play—whether isolated or social, informal or planned, analogue or digital—are more or less linked to opportunities for creative thought and action. It would appear, perhaps, that all advocates of 'free' play time and space for children are aiming to increase children's creativity.

But not all those who champion play do so in ways that are conducive to the freedom of thought, creative action or divergent and critical thinking that are suggested in other rhetorics as being the prime ingredients of creativity. Nor, indeed, as Shanly Dixon and Sandra Weber's chapter 'Play Spaces, Childhood and Video games' (2007) shows, are all rhetorics of *Play and Creativity* motivated by the same goals and histories. They continue important discussions (Sutton-Smith, 2001) about the links between adults' nostalgia for a remembered context of play in their own childhoods and emerging, ingrained and often naturalised social rhetorics about play in modern children's lives. Taking to task those who mourn the 'death' of an era when play was outdoors, safe, free and unmediated, they write:

> In response to both panic and nostalgia, adults are increasingly organizing and regulating their children's play. Contemporary childhood is

Figure 9.1   3-year-olds making shapes with modeling dough. Photograph by the author.

now constructed by adults as a space where children must continuously be engaged in activities that are productive. There is an expectation that play must serve some higher purpose: for instance, children play to learn, children play to burn off excess energy, children play for exercise. Play is no longer an objective in and of itself (Sutton-Smith, 1997). As a result . . . free play is becoming an oxymoron rather than a logical coupling of complementary words. (2007, p. 25)

In quite specific ways this discussion can be seen to mirror discourses that have emerged with regard to creativity, technology and (new) media. Cordes and Miller, for instance, assert that

[c]reativity and imagination are prerequisites for innovative thinking, which will never be obsolete in the workplace. Yet a heavy diet of ready-made computer images and programmed toys appear to stunt imaginative thinking. Teachers report that children in our electronic

society are becoming alarmingly deficient in generating their own images and ideas. (2000, p. 4)

But the fact that certain commentators, possibly with nostalgic memories of socially privileged childhoods and an exaggerated paranoia about 'modern' media, might overstate the case against digital playtime does not mean that all technology-based play and learning are either harmful or necessarily beneficial to children's creativity.

A DIGITAL 'CREATIVITY PILL' OR A DAMAGING POTION?

If creativity is not inherent in human mental powers and is, in fact, social and situational, then technological developments may well be linked to advances in the creativity of individual users. The rhetoric constructed around *The Creative Affordances of Technology* covers a range of positions from those who applaud all technology as inherently creative to those that welcome it cautiously and see creativity as residing in an as yet under-theorised relationship between users and applications. But it is worth asking how democratic notions of creativity are linked to technological change in this rhetoric. Is the use of technology itself inherently creative? And how do concerns raised by opponents of new technology affect arguments about creative production?

For Avril Loveless (2002), because of a complex set of features of ICT (provisionality, interactivity, capacity, range, speed and automatic functions), digital technologies open up new and authentic ways of being creative 'in ways which have not been as accessible or immediate without new technologies' (2002, p. 2). Loveless (1999) explores some of the issues arising with regard to visual literacy and multimedia work for classroom teachers. She notes that during the Glebe School project, in addition to the generation of great enthusiasm and enjoyment during the use of visual packages on the computer, the question of evaluation was not forgotten by the children who 'had a sense of ownership of their images and lively ideas on how they might adapt or improve them in the future' (1999, p. 38). Viewing their digital media projects, the children felt that the finished pieces did not look like 'children's work' (1999, p. 39), and would hence be taken more seriously by adults evaluating and appreciating them. Teachers however expressed a variety of concerns about the potentials and actuality of such ICT-related projects for their students and themselves. They were concerned about their own levels of understanding and skill in relation to the software. Given this context, Loveless argues that technology, which is now being used in schools in varieties of ways, can enhance creative learning, but only if teachers' anxieties are handled sensitively via 'the *strategic* approach to the use of ICT' through the application of skills, 'rather than skills training associated with specific packages' (1999, p. 40, emphasis in original).

Implicitly addressing many of the concerns aired in educational circles (Healy, 1998; Reid, Burn, & Parker, 2002) about the apparently empty 'showiness' of digital products by children, Loveless (2003, pp. 13–14) cautions against using tools and techniques in digital creations so that one can say that children's work is technology-enhanced. Regardless of the technological tools being used by the children in their work, she suggests that it is the meanings being produced by the children in their projects, the children's and teachers' sense of confidence with and ownership of their tools, and ongoing pedagogic negotiations that will aid or hinder creative production. Supporting this socially situated view of the *potentially* creative uses of digital technologies in their riposte to one particularly wide-ranging and trenchant critique (Cordes & Miller, 2000), Douglas Clements and Julie Sarama (2003) cite studies that document what they call 'increases in creativity' and as well as better peer relations following interactive experiences with computer programmes such as Logo. However, challenging those who champion digital technologies as *inherently* creative, like Scanlon, Buckingham, and Burn (2005) and Seiter (2005), they also note that many computer programmes designed to increase children's knowledge and skills are not in the least bit creative, relying on rote learning, repetition and drill exercises. Thus they argue that digital technology can, but does not necessarily, support the expression and development of creativity.

Sefton-Green (1999, pp. 146–147) notes that successful digital projects with children and youth are all heavily intensive in terms of time, staff and resources. Here, despite the enthusiasm generated, 'the organisation of the school day with its narrow subject disciplines, short working periods, and heavy assessment load' (Sefton-Green, 1999: 146–147) are seen as opposed to the principles of digital arts work and as inhibiting the success of such projects particularly in secondary school. Furthermore, the projects in Sefton-Green's collection all raise significant questions about the evaluation of creative work in new media more generally: 'Do we evaluate students' grasp of authoring packages or their capacity to imagine in the new medium?', 'When is a product genuinely interactive and when does it merely ape fashionable conventions?' (1999, p. 149). In a society where technology is not equally available to all, children may well be enthusiastic and confident users of digital technologies when offered the opportunities for playful production, but they are still divided by inequalities of access outside school and across the school system (1999, p. 153). Ultimately the social contexts of digital technology's use may help or hinder its creative potential.

EVALUATION, LEARNING AND PEDAGOGY

Pertinently for those interested in creativity and communication, placing itself squarely at the heart of educational practice, *The Creative Classroom* rhetoric focuses on pedagogy, investigating questions about the connections between knowledge, skills, literacy, teaching and learning and the place of creativity in an increasingly regulated and monitored curriculum (cf. Beetlestone, 1998; Starko, 2005; Jeffrey, 2005). The focal point of this rhetoric is frequently practical advice to educators in both formal and informal settings about ways of encouraging and improving the learning of young people. This rhetoric locates itself in pragmatic accounts of 'the craft of the classroom', rather than in academic theories of mind or culture. Creative learning is *interactive*, incorporating discussion, social context, sensitivity to others, the acquisition and improvement of literacy skills; it is *contextual*, and has a sense of *purpose* and thus cannot be based around small units of testable knowledge; however, it can also be thematic and highly specific as it often arises out of stories or close observation, which engage the imagination and the emotions as well as learners' curiosity about concepts and situations.

In this view, in terms of the content of creative lessons, it is vital that concepts are not taught as being fixed and immutable entities but as contextually and culturally anchored; subject divisions too need to be seen as frequently arbitrary and socially constructed rather than as rigid and binding; for it is in crossing such divisions—between art and mathematics, physical activity, numeracy, languages and music, geography and science, philosophy and poetry that children (and adults) stand the greatest chance of being independently creative. All this is unquestionably sound advice. Indeed, the *Creative Classroom* rhetoric is consistent in identifying holistic teaching and learning—which link playful processes to different types and domains of knowledge and methods of communication—as more compatible with and conducive to creative thought and production than the increasingly splintered, decontextualised, top-down and monitored content and skills which are favoured as being academically 'effective'. There is, however, a tension in this work between what could broadly be defined as a rather romantic wish to view creativity as something that enhances the human soul and helps young people to blossom, and the need to give practical advice to trainee teachers, thus fitting them for the fairly chaotic but restricted milieu into which they will soon be going. At points this tension is productive, or at the very least practical, in the sense that it prevents the educational perspective on creativity from sidestepping issues such as assessment and time management that are of very real significance for practitioners both in formal educational and more unorthodox settings.

The examples of 'creative teaching' given exemplify the tightrope that many educators have to walk between institutional constraints and the fragility of their constructed 'creative' environment. However, at times the tension also appears to lead to contradiction or even paradox: risk-taking is to be encouraged but it is also to be kept within easily controllable bounds; time is required for playful engagement with ideas and materials, but this time has stringent external parameters in terms of the school day. Alpesh

Maisuria argues powerfully that the interventions of recent governments in education have created a culture of 'vocationalisation', 'standardization' and 'rubber stamp' testing which has all but killed the space for creative pedagogy, playful exploration and creative work in the classroom:

> Performance indicators and standards inspectorates culminate in teachers avoiding risks (Campbell, 1998). Teachers are positioned in a catch-22 situation where they are inclined to conform to the curriculum specification rather than indulge in vibrant and energised pedagogy driven by ingenuity. Teachers do not encourage independent thinking and elaborate innovations because the curriculum and standards criteria do not recognise unorthodox creative expressionism (Davis, 2000). (Maisuria, 2005, p. 143)

While it is clear that a number of students continue to work in imaginative and divergent ways, and that some teachers still encourage them to do so by valuing playful or subversive discussion and creative production with new or traditional technologies, the literature on creativity in contemporary classroom settings suggests that this is despite, rather than because of, current education policies. In their study of the ways in which university staff and students experience and understand creativity, Oliver, Shah, McGoldrick, and Edwards' (2006) interviews uncover a number of experiences and patterns that fit in with the rhetorics outlined so far: a liking for creative or inspirational teachers/lecturers and a sense that being around enthusiastic, critical and engaged individuals enhances creativity; a dislike of dogmatic teaching, deadlines, narrow theoretical parameters, subject hierarchies which devalue drama and the arts in relation to mathematics and science; depression at the lack of reward for critical or divergent work and about forced targets; as well as anxieties around performing creativity 'on demand' and being shown as uncreative in front of other students were frequently expressed. However, highlighting institutional barriers to individual and group creativity, 'in many students' comments there was a sense of frustration at a perceived conflict between being creative and being "academic"' (2006, p. 54).

Although not considered in detail here, in response to such institutional realities, and setting a challenge to aspects of foregoing rhetorics, *Creative Arts and Political Challenge* sees art and participation in creative education as necessarily politically challenging, and potentially transformative of the consciousness of those who engage in it. It describes the processes of institutional pressure that militate against positive and challenging experiences of creativity by young people, regardless of the efforts of teachers and practitioners (Thomson, Hall, & Russell, 2006). In previous work on this topic (Banaji and Burn 2007a; Banaji & Burn 2007b), this rhetoric is pursued further, with an emphasis on questions it raises about creative partnerships, social contexts and political or philosophical presuppositions. If one wishes to retain the idea of cultural creativity as having an oppositional rather than a merely socialising force, it is important not to lose sight of the ways in which broader inflections of discourses of creativity relate to the micropolitics of particular social settings. The very fluidity and confusion in talk about creativity in the classroom can mean that the term is used as window dressing to appease educators who are interested in child-centred learning without actually being incorporated into the substantive work of the classroom.

CONCLUSION

In exploring questions about the nature and significance of creativity via engagement with symptomatic texts that use one or more of the different rhetorics this chapter has raised a number of issues. The public discourse on creativity is characterised by a lack of clarity that allows participants to gain the benefits of aligning themselves with conflicting or mutually incompatible ideas and views without being seen to do so. In the 1990s the rush to install computers in schools apparently to aid children's digital creativity and their preparedness for a modern economy has been accompanied by hysteria about how computer use impairs traditional literacy and creativity. Similarly, various proponents of creative arts in the classroom have claimed for arts projects a huge democratising effect while in practice holding firmly elitist beliefs about artistic and literary products. One of the dangers of purely cognitivist conceptions of creativity is that they lose a sense of cultural groundedness, provenance, and of the cultural experiences of learners prior to any given educational experience, whether within or beyond formal education. In an educational context, the emphasis on creativity is part of an effort to draw back from the perceived excesses of a highly regulated, performance-based audit culture and to recover something that existed before, whether this be called 'enjoyment', 'good teaching' or 'creativity' without, however, losing apparent 'excellence' and 'standards'. Unfortunately, given that currently 'excellence' and 'standards' are criteria that are set by the very 'audit culture' from which *The Creative Classroom* and *Play and Creativity* rhetorics hope to depart, there is a significant issue here in terms of the emphasis which is given by those carrying out assessments to processes of learning or creating and the products or the absence thereof. We are left with the question: Is play uncreative if it does not produce a tangible product?

Another strong strand identified in this chapter is a relatively bland discourse of prosocial intervention: creative projects and strategies that encourage tolerance, cooperation and social harmony. A sharper version of that argument posits creativity as being about social inclusion and cultural diversity. In the name of creativity, this rhetoric uses broad aspirational terms such as empowerment and democratisation, although the precise nature of the goals that are sought often remains unclear. But assessing whether any of the grand or even the more modest political ambitions of particular rhetorics and creative projects have been achieved is not easy. How do we assess whether or not children have been empowered or local communities made more tolerant or workforces made more collaborative? It is crucial that we understand and respond to the relationship between the *cultural politics* of talk about creativity or play and a *wider politics*. While there is evidence from numerous studies (Balshaw, 2004; Starko, 2005) that creative ways of teaching and learning, and creative projects in the arts, humanities and the sciences, offer a wider range of learners a more enjoyable, flexible and independent experience of education than some traditional methods, there is no evidence that simply giving young people or workers brief opportunities for creative play or work substantially alters social inequalities, exclusions and injustices. Creativity is not a substitute for social justice.

There is a complex and not always clearly identifiable *cultural politics* behind many rhetorics of creativity as there is behind educational rhetorics and the rhetorics of play. This is the case not only within discourses which explicitly address questions about power, and about whose culture is seen as legitimate and whose is not. It is also the case in discourses where constructions of power remain implicit, such as those which celebrate 'high art' as 'civilising', child art as being about an 'expression of the soul', or which see the development of workers' creativity as being 'for the good of the national economy' and a constant testing and attributions of levels of ability to children as a way of raising 'standards'. The word 'productive', when used in relation to children's play, is especially poignant in terms of discursive constructions of creativity and the social structures which inform them. It belies all the supposed emphasis on 'freedom' and 'agency' in discussions of childhoods past and present. As may be observed in the case studies included in this volume, most children do not stop after playing to measure the quality of their playtime by analysing its physical products (drawings, paper planes) or other effects (tiredness, relaxation, amusement). The suspicion evinced by some parents and educators with regard to the amount of time children spend watching television, reading blogs or playing computer games rather than reading or playing cricket can be seen to spring from complex social discourses about 'healthy' play and 'harmful' play, about what is recognized as creative versus what is labeled as 'derivative' and about what children want to do in their spare time. Less significant than the specifics of what aspect of creativity is sanctioned and what is not at any given historical moment, is the fact that some rhetorics explicitly legitimise certain forms of cultural expression and certain goals and implicitly delegitimise others. Whether the labels 'digital' and 'creative' are applied pejoratively or to applaud, some contemporary rhetorics can and do aid social gatekeeping by stigmatising particular pedagogies and parenting choices. Rhetorics of creativity are, then, always political, even when they appear not to be.

ACKNOWLEDGMENTS

In formulating the rhetorics that appear here and in tracing their lineage, I am grateful for the substantial contributions and critiques of Andrew Burn and David Buckingham at the Centre for the Study of Children, Youth and Media. I also thank Creative Partnerships for the opportunity to research and write the literature review from which this chapter arises and the Arts Council for permission to reproduce a section of that literature review.

NOTES

1. See, for instance, the websites for *Creative Thinking and Lateral Thinking Techniques* (2003), available at http://www.brainstorming.co.uk/tutorials/creativethinkingcontents.html, and *Creativity Unleashed Limited* (2003), managerial training website at http://www.cul.co.uk/

REFERENCES

Banaji, S., & Burn A. (2007a). Creativity through a rhetorical lens: implications for schooling, literacy and media education. *Creativity and Literacy, Literacies, 41*(2), 62–70.

Banaji, S., & Burn, A. (2007b). *The rhetorics of creativity: A review of the literature.* London: Arts Council of England.

Balshaw, M. (2004). Risking creativity: Building the creative context. *Support for Learning, 19*(2), 71–76

Beetlestone, F. (1998). *Creative children, imaginative teaching.* Buckingham, UK: Open University Press.

Brennan, C. (2004) (Ed.). *Power of Play: A Play Curriculum in Action Play.* Dublin: IPPA.

———. (2005). Supporting Play, Supporting Quality Conference Proceedings, *Questions of Quality*: September 23–25, 2004, Centre for Early Childhood Development and Education, Dublin Castle, 302–311.

Boden, M. (1990). *The creative mind: Myths and mechanisms.* London: Weidenfeld and Nicolson.

Buckingham, D., & Jones, K. (2001). New Labour's cultural turn: Some tensions in contemporary educational and cultural policy. *Journal of Educational Policy, 16*(1), 1–14

Campbell, J. (1998). Primary considerations: Broader thinking about the primary school curriculum. In S. Dainton (Ed.), *Take care, Mr Blunkett: Powerful voices in the new curriculum debate* (pp. 96–100). London: Association of Teachers and Lecturers.

Carruthers, P. (2002). Human creativity: Its cognitive basis, its evolution, and its connection with childhood pretence. *British Journal for the Philosophy of Science, 53*, 2: 225–249.

Clements, D. H., & Sarama, J. (2003). *Strip mining for gold: Research and policy in educational technology—A response to "Fool's Gold." Association for the Education of Young Children Journal,* 11(1), 7–69.

Cohen, G. (2000). *The creative age: Awakening human potential in the second half of life.* New York: HarperCollins.

Cordes, C., & Miller, E. (2000). *Fool's gold: A critical look at computers in childhood.* Alliance for Childhood. Retrieved July 7, 2007, from http://www.allianceforchildhood.net/projects/computers/computers_reports_fools_gold_download.htm

Craft, A. (1999). *Teaching creativity: Philosophy and practice.* London and New York: Routledge.

Craft, A. (2003). Creative thinking in the early years of education. *Early Years,* 23(2), 147–158.

Creative Thinking and Lateral Thinking Techniques. (2003). Retrieved June 29, 2007, from http://www.brainstorming.co.uk/tutorials/creativethinkingcontents.html

Creativity Unleashed Limited. (2003). Retrieved June 29, 2007, from http://www.cul.co.uk/

Cremin, T., Comber, B., & Wolf, S. (Eds.). (2007, July). *Creativity and Literacy,* Special Issue editorial *Literacies,* 41(2): 55-61.

Cropley, A. J. (2001). *Creativity in education and learning: A guide for teachers and educators.* London: Kogan Page.

Csikszentmihalyi, M. (1997). *Creativity: Flow and the psychology of discovery and invention.* New York: Harper Perennial.

Cunningham, H. (1998). Digital culture—The view from the dance floor. In J. Sefton-Green (Ed.), *Digital diversions: Youth culture in the age of multimedia* (pp. 128–148). London and Bristol, PA: UCL Press.

Davis, T. (2000). Confidence! Its role in the creative teaching and learning of design and technology. *Journal of Technology Education,* 12(1), 18–31.

Dixon, S., & Weber, S. (2007). Play spaces, childhood and video games. In S. Weber & S. Dixon, (Eds.), *Growing up online: Young people's everyday use of digital technologies* (pp. 15–33). New York: Palgrave MacMillan.

Gardner, H. (1993). *Frames of mind: The theory of multiple intelligences.* London: Fontana Press.

Gauntlett. D. (2007). *Creative explorations: New approaches to identities and audiences.* London and New York: Routledge.

Healy, J. (1998). *Failure to connect: How computers affect our children's minds—for better and worse.* New York: Simon & Schuster.

Illinois loop. (2007). Retrieved March 28, 2007 from http://www.illinoisloop.org/artmusic.html

Jeffrey, G. (Ed.). (2005). *The creative college: Building a successful learning culture in the arts.* Stoke on Trent, UK and Sterling and Sterling, USA: Trentham Books.

Kant, I. (2000). *The critique of judgement.* New York: Prometheus Books. (Original work published 1790)

Kwek, D., Albright, J., & Kramer-Dahl, A. (2007). Building teachers' creative capabilities in Singapore's English classrooms: A way of contesting pedagogical instrumentality. In *Creativity and Literacy, Literacies,* 41(2), 71–78.

Landry, C. (2000). *The creative city: A toolkit for urban innovators.* London, UK, and Sterling, USA: Commedia, Earthscan Publications.

Loveless, A. M. (1999). A digital big breakfast: The Glebe School Project. In J. Sefton-Green, (Ed.), *Young people, creativity and new technologies: The challenge of digital arts* (pp 32–41). London and New York: Routledge.

Loveless, A. M. (2002). *Literature review in creativity, new technologies and learning.* NESTA Futurelab.

Loveless, A. M. (2003). Creating spaces in the primary curriculum: ICT in creative subjects. *The Curriculum Journal, 14*(1) 5–21.

Marshall, B. (2001). Creating danger: The place of the arts in education policy. *Creativity in education* Craft, A., Jeffrey, B. and Liebling, M. (Eds.) London: Continuum, pp. 116–125.

Maisuria, A. (2005). The turbulent times of creativity in the national curriculum. *Policy Futures in Education, 3*(2), 141–152.

Negus, K., & Pickering, M. (2004). *Creativity, communication and cultural value.* London, Thousand Oaks, CA, and New Delhi: SAGE.

Oliver, M.; Shah, B., McGoldrick, C., & Edwards, M. (2006). Students' experiences of creativity. In N. Jackson, M. Oliver, M. Shaw, & J. Wisdom, (Eds.), *Developing creativity in higher education: An imaginative curriculum* (pp. 43–58). London and New York: Taylor and Francis.

Osborn-Parnes–1941 (This is a model, not a book) I.e. The Osborn-Parnes model.

Pope, R. (2005). *Creativity: Theory, history, practice.* London and New York: Routledge.

Reid, M., Burn, A., & Parker, D. (2002). Evaluation report of the BECTa digital

video pilot project. Retrieved July 7, 2007, from http://www.becta.org.uk/research/reports/digitalvideo/index.html

Robinson, K., et al., & National Advisory Committee on Creative and Cultural Education. (1999). *All our futures: Creativity, culture and education.* Sudbury, Suffolk, DfEE publications, NACCCE, 220.

Russ, S. (2003). Play and creativity: Developmental issues. *Scandinavian Journal of Educational Research, 47*(3), 291–303.

Scanlon, M., Buckingham, D., & Burn, A. (2005). Motivating maths: Digital games and mathematical learning. *Technology, Pedagogy and Education, 14*(1), 127–139.

Scholtz, A., & Livingstone, D. W. (2005). 'Knowledge workers' and the 'new economy' in Canada: 1983–2004. Paper presented at 3rd annual Work and Life Long Learning (WALL) conference, Toronto, Canada.

Scruton, R. (2000). After modernism. *City Journal, 10*(2): NP. Available at http://www.city-journal.org/html/10_2_urbanities-after_modernis.html. Last accessed 25th June 2008.

Sefton-Green, J. (1999). A framework for digital arts and the curriculum. In J. Sefton-Green (Ed.), *Young people, creativity and new technologies: The challenge of digital arts* (pp 146–154). London and New York: Routledge.

Seiter, E. (2005). *The Internet playground: Children's access, entertainment and mis-education.* New York: Peter Lang.

Seltzer, K., & Bentley, T. (1999). *The creative age: Knowledge and skills for the new economy.* London, UK: DEMOS.

Simonton, D. K. (1999). Genius, creativity, and leadership: Historiometric inquiries. Harvard, USA: Harvard University Press.

Starko, A. J. (2005). *Creativity in the classroom: Schools of curious delight.* New Jersey, USA: Lawrence Erlbaum Pubishers.

Sutton-Smith, B. (2001). *The ambiguity of play.* Boston: Harvard University Press.

Thomson, P., Hall, C., & Russell, L. (2006). An arts project failed, censored or . . . ? A critical incident approach to artist-school partnerships. *Changing English,* Volume 13, Number 1, April 2006, 29–44.

Vygotsky, L. S. (1994). Imagination and creativity in the adolescent. In R. Van Der Veer & J. Valsinger (Eds.), *The Vygotsky Reader* (pp. 288). Oxford, UK, and Cambridge, MA: Blackwell. (Original work published 1931)

Willis, P. (1990). *Common culture.* Milton Keynes, UK: Open University Press.

10 What Education Has to Teach Us About Games and Game Play

Caroline Pelletier

Debates about the educational value of computer games have tended to focus on how education can benefit from their informational and skill-based content or their embedded pedagogy. This debate often relies on a restricted conception of play, which emphasises its benevolent and self-motivating character at the expense of its role in hierarchical peer relations, power negotiations and institutionalised rituals. In this chapter, I examine certain conceptions of play in education and then go on to study the social functions of game play in peer and power relations in an after-school club run by researchers in London between 2004 and 2006. The aim of the club was to enable 12–14-year-olds draw on their experience of game play to create their own computer games. The rationale for this endeavour emerged from a media education tradition (Buckingham, 2003; Burn & Durran, 2007; Hobbs, 1998). This emphasises the importance of extending conceptions of literacy to multimedia texts, and developing communicative abilities in a variety of modes and genres (see Carrington, this volume). My focus is on how conceptions and enactments of what 'play' entails emerged from experiences of game play in the club, including social relations between club members. A second point of focus will be on the approach taken by researchers to teach about game design. These two points of interest are linked in that I will trace students' conceptions and enactments of play over a number of weeks to comment on the potential usefulness of the approach taken in the club to develop a more complex understanding of games and game play.

CONCEPTIONS OF PLAY AND THEIR IMPLICATIONS FOR EDUCATION

A common way of defining play is that it is the opposite of work. On the basis of this definition, play encompasses a wide range of activities. Such activities are said to be playful to the extent that they are intrinsically pleasurable, internally motivated, self-contained and surplus to utilitarian goals. Roger Caillois (1967), for example, defines play as something which is essentially gratuitous, voluntary and disinterested, a spontaneous instinct based on freedom and fantasy, opposed to work in that it does not produce anything (it is noninstrumental). Karl Groos (1898/2005) and Johan Huizinga (1938/1970), early theorists of play, similarly emphasise that play cannot be imposed by physical or moral necessity; play is that which escapes the rigours of everyday existence.

According to Sutton-Smith (1997), it is this definition of play which, in education, underpins efforts to recuperate play by making it useful with respect to the development of adult skills; in other words, play is defined as intrinsically motivating precisely in order to sustain the belief that play is an evolutionary mechanism whose value is realised later in life. Farley (2006) makes a similar point in suggesting that it is precisely when play is constructed as intrinsically motivating (i.e., nonutilitarian) that it is understood to serve the functional ends of educational progress. Play is redeemed as an educationally useful activity by being framed as a safe zone in which children can deal with experience by creating model situations and mastering reality through experimentation. Such a view of play has antecedents in Freud's (2003) work on play as therapy, and can be found in Piaget's (1965/1976) argument about the role of play in the child's moral and cognitive development, as well as Bruner's (1976) emphasis on play as a form of problem solving.

This view of play also characterises some recent arguments about the educational value of computer games. In Gee's work (2003, 2004), computer game play is said to establish 'a *psychological moratorium* [...] in which the learner can take risks where real-world consequences are lowered' (2003, p. 62, original emphasis). Game play sustains 'affinity spaces' which involve 'porous leadership' and leaders who 'don't and can't order people around' (2004, p. 87; [see Whiteman (2007) for a study which challenges Gee's view that game-based affinity groups are removed from the hierarchies and conflicts of interest of the 'real world']). The BECTA report (Dawes & Dumbleton 2001, p. 2) on computer games similarly defines their educational potential in terms of 'that elusive and invaluable characteristic—motivation', a view which frames game play as intrinsically motivating and freely engaged with, and precisely in this respect distinct from school-based learning.

Sutton-Smith (1997) critiques a view of play as intrinsically motivating and inherently benevolent in terms of a rhetoric of 'play as progress'. The term *rhetoric* is used in the same way as in Banaji's chapter in this volume, to denote the way in which a phenomenon is conceptualised and the ideological underpinnings of such a conception. Sutton-Smith does not argue that such a rhetoric is wrong as such, but that it neglects the way in which play is a practice of communication and representation, which has real consequences for players, in the present rather than merely in an adult future. This rhetoric is an ideological strategy, he suggests, to constrain children's play in the service of growth and education. It works by differentiating between creative, progressive playfulness (that of the rich) and idle, unproductive, depraved play (that of the poor; Sutton-Smith, 1982, p. 195).

Sutton-Smith's own work focuses on the function of play in community building: 'the major meaning of social play that emerges from a review of folkloric material is that play is about power and the struggle for identity within the dominance–subordination domains of one's peers' (1984, p. 61). This argument highlights the role which play has in forming and negotiating social bonds: 'A feature that is almost unanimously acknowledged to be the hallmark of play is that it is intrinsically motivated (that it is "fun"). . . . Many anthropologists and historians would have trouble with this account since most play throughout the ages has been extrinsically motivated by village requirements' (1997, p. 188): In other words, the requirement to be part of a community or cultural group. Such a requirement need not necessarily be understood as 'extrinsically motivated', but can be conceptualised as an internalised desire for cultural belonging. Rather than being removed from real life, then, secluded within a magic circle, play is ritualistic, concerned with establishing and maintaining social norms. This is not to deny the phantasmagorical or pleasurable dimension of play. But it is to frame play as a social practice, embedded in social rituals, with its own instrumental behaviours. Sutton-Smith here makes a similar argument to Ehrmann (1968), who critiques Huizinga and Caillois by highlighting the role of play in acquiring prestige and power in communities.

Sutton-Smith's and Ehrmann's anthropological perspectives complement a media education argument about the value of studying computer games, in that game play is conceptualised as a significant social activity, rather than primarily in terms of cognitive adaptation or as a moratorium outside social processes. Game play is relevant to education as a practice of social exchange and identity formation, and so warrants understanding in its own right.

A second but related conception of play in the field of education emerges from the grammatical necessity which links 'game' to 'play': Games are played, they are not 'done' or 'performed'. Consequently, any interaction with a game can be identified as play. This grammatical necessity seems to underpin some of the recent interest in the formal qualities of games: their components, design patterns, the nature of their interfaces. Kirriemuir and McFarlane (2004) argue that computer games interest educationalists because of their 'desire to harness the motivational power of games' (p. 4). This desire sustains support for the development of educational games: software that looks like a game, has many components of a game (objectives, characters, animated colourful graphics, feedback loops, etc.), but which has educational content (see, e.g., DfES 2003, 2005 for recommendations about the development of games matched to curriculum objectives). This line of argument is based on a metaphor: If it looks like a game, has the same formal design attributes as a game, it will be *like* a game (with game defined as motivating, fun, playful).

Classifying types of games and identifying design patterns informs the study of games. However, the effort to identify design patterns in order to make educational software which is playful—understood as motivating, fun—is based on an assumed symmetry between form and meaning which transcends context. Whilst acknowledging that games are played, are they also playful irrespective of differences across time and space? The BECTA report suggests that games like *The Sims* or *Championship Manager* could teach skills such as budget management and database handling in a more motivating way (Dawes & Dumbleton, 2001). But would *The Sims* remain playful as part of a curriculum on household budget management? Bateson's (1955/1976) definition of play as a form of metacommunication suggests not. In his famous formulation of the paradox of play, he states: 'a playful nip denotes the bite, but it does not denote what would be denoted by the bite' (p. 121). Play is not a particular kind of content but the meaning of that content within a specific context. Bateson's formulation ironically suggests that *The Sims* is playful precisely to the extent that it is *not* concerned with the serious business of household budget management. Games are pleasurable, according to Bateson, because of the meaning players attribute to their actions, rather than as a result of the content or formal arrangement of play.

Bateson's formulation again suggests an argument for studying games which is not primarily about making content more motivating. Rather, it frames game play as a signifying practice. The educational value of games can consequently be understood in terms of their significance in social and cultural practices.

THE 'MAKING GAMES' PROJECT

It was on the basis of computer games' cultural significance that the research project 'Making Games' was set up at the Institute of Education, London. It ran for 3 years and developed approaches to the study of games in formal and informal contexts, within a media education framework.[1] The rationale for the project related to the significance of games as a genre, and to the importance of enabling students develop communicative abilities in this genre, by both analysing and making games. The creation of software which enables young people to produce games was central to the research, by facilitating the integration of critical analysis and creative expression (Buckingham, Grahame, & Sefton-Green, 1995).

In the remainder of this chapter, I present one approach we used for studying games in an after-school club. Variations on this approach were subsequently developed and included in a teacher support pack for secondary level English and Media teachers.[2] My description is organised to focus on students' emerging conceptions of play. This highlights the relationship between the pedagogy, identity development and conceptions of game play.

The after-school club involved a core group of nine students and took place in an inner city school in London. It ran over the length of one term every year for 3 years, and consisted of one hour sessions each week. Students were approached individually by teachers although participation was voluntary.

The approach taken in the after-school club was informed by prior fieldwork in a school in Cambridge, where researchers contributed to the design of a Year 8 English, Media and ICT course about computer games. For this course, we adapted an established model in media education, looking at games in terms of text (as representation), institutions and audiences (Buckingham, 2003). At the end of the course, students' paper-based designs and conceptions of game design and play reflected the emphasis placed on representation, with most attention given to narrative and visual imagery. We found this problematic, precisely because earlier work on the textuality of games emphasised that games operate at two levels: the representational, but also the ludic (Carr, Buckingham, Burn, & Schott, 2006). The field work suggested that the traditional emphasis on representation in media education did not give sufficient attention to games as systems for interaction, or mechanics of play.

Consequently, we decided to focus on developing a conception of games as interactive systems that sustain play. To achieve this, we used board games, for three reasons. First, as Salen and Zimmerman (2003) argue, all games can be conceptualised in terms of a single field of design; The same design principles underpin games from 'Spin the Bottle' to *World of Warcraft*. Second, board games are more easily manipulable than computer games: Their components are material, which means they can be more easily handled, adapted, edited and tested than software. Third, initial fieldwork suggested that young people had a limited vocabulary for describing games; they could not articulate, and thus identify, what games consisted of in terms of ludic systems, why some games were better than others, or which aspects of games they would like to incorporate into their own designs. This reflects widespread limitations in understanding about games, which are now addressed in the field of 'game studies'. However, to develop a precise and systematic vocabulary for talking about games, we wanted to integrate analysis and play. Board games can, in an after-school club, be played collectively and interrupted for analysis more easily than computer games.

THE BOARD GAME ANALYSIS COURSE

In order to better understand the relationship between the design of games and how they were played, we played a number of widely known games: Snakes & Ladders, Chess, Monopoly and Cluedo. In each session we played the games as a group, then paused play to address a number of questions, such as:

- What are the rules of the game? What role do different rules have? How do you learn the rules? How do you develop a playing strategy? What is the core mechanic of each game?
- What are the components of the game and how do they relate to each other?
- How are the rules and the narrative setting integrated?

Students were then asked to change the rules of the game:

- How can Snakes & Ladders be made into a more strategic game? How can chance be introduced into Chess?
- How can the rules of Monopoly be altered to make the game finish more quickly, or go on for even longer? What happens to game play if the value of certain squares in Monopoly are altered? What happens if players start with less/more money?
- How could the rules of Cluedo be combined with those of Monopoly or Snakes & Ladders? How could Cluedo be adapted to a computer game format?

This approach was informed by Salen and Zimmerman (2003) who suggest ways of analysing games as ludic systems.

In the following two sections, I analyse transcripts from two sessions, 3 weeks apart. My interest here is in highlighting the relationship between conceptions of play, play behaviour and the formation of social relations.

PLAYING AND ANALYSING THE GAME OF SNAKES & LADDERS

The discussion below involved seven students, one teacher (T) and two researchers, Andrew Burn and myself. Andrew and I here behaved like teachers. At the start of the session, six students were asked to play Snakes & Ladders, with one asking to play Chess with the teacher. Andrew and I sat among the students but did not play. Activities were captured on a fixed video camera. After 15 minutes, Andrew brought play to a halt:

1.	*Andrew*	Stop, Stop. Right, well we'll talk about Snakes & Ladders first
2.		alright? So, lets just go round, alright, everyone say one thing
3.		about Snakes & Ladders, to tell us. If I was the Martian, just
4.		landing on earth and I wander up to the two of you and you are
5.		explaining to me, alright?
6.	*Jameela*	You have to go up the ladders and down the snakes
7.	*Michelle*	Erm . . . you roll dice [laughter]
8.	*Andrew*	Dice?
9.	*Michelle*	Oh, it's a little cube thing with loads of numbers on it between

10.		one and six
11.	*Andrew*	Thank you
12.	*Helen*	[pause as Helen hesitates—general laughter]
13.		Erm . . . you have to take turns to go around
14.	*Andrew*	OK
15.	*Alisha*	Takes several people
16.	*Helen*	Yeah
17.	*Jameela*	I think you can play with one person yeah?
18.	*Alisha*	But it's better with two people or more
19.	*Andrew*	OK
20.	*Caroline*	It would be a bit dull if it was you just on your own, wouldn't
21.		it?
22.	*Andrew*	Yes it would be wouldn't it. Well, you know, solitaire snakes
23.		and ladders
24.		[laughter]
25.	*Moire*	You use markers to show where you are
26.	*Caroline*	Yeah
27.	*Andrew*	OK
28.	*Vanessa*	I can't think of anything else
29.	*Andrew*	OK. So what are the rules, or are there rules?
30.	*Jameela*	You have to get a six before you can move
31.	*Andrew*	You have to get a six before you can move. But you, you
32.		had another rule about sixes, didn't you?
33.	*Michelle*	Yeah, sort of
34.	*Jameela*	If you throw a six, you get another go
35.	*Andrew*	But you only just decided on this half way. Is this the rule? Or
36.		are they just making this up?
37.	*Michelle*	Yeah, that's the rule
38.	*Andrew*	That's a rule. You agree with that. Excellent. No? [Michelle shakes head,
39.		Andrew laughs]. OK any other rules about snakes and ladders?
40.	*T*	Can you go up the snake?
41.	*General*	No
42.	*Alisha*	I think we should add a rule where if you are sitting at the top of
43.		the ladder, you have to go down it
44.	*Andrew*	OK
45.	*Caroline*	And then go up the snake?
46.	*Alisha*	Erm . . . [laughter]
47.	*Michelle*	Yeah, they are longer, the snakes are longer, you could go up
48.		them instead
49.	*Helen*	You go up the yellow snakes and down the blue snakes. And
50.		then you could go to the bottom of the ladder
51.	*Alisha*	What about the green snakes?
52.	*General*	[laughter]
53.	*Helen*	Are there green snakes? Just one
54.	*Alisha*	One

55.	*Michelle*	I love those green ones as well, they are nice
56.	*Alisha*	What do you say?
57.	*Caroline*	So how do you win in Snakes and Ladders?
58.	*Andrew*	Yeah
59.	*General*	When you get to the last square
60.	*Caroline*	And how do you get to the finish?
61.	*Michelle*	By going there, so by moving these [moves counter along,
62.		banging it on table]
63.	*Caroline*	OK, well you have to land on one hundred exactly, don't you?
64.	*General*	Yeah
65.	*Jameela*	If you go over, it's like . . . go five, one, to, three, four, five
66.		[moves counter along to one hundred and then back again. Last
67.		square has a snake on it, so Jameela slides counter down it]
68.	*Caroline*	Ah, so that snake is a real . . . that snake is awkward
69.	*Alisha*	No, you weren't there [argument about where counters were
70.		originally before the game was interrupted—aimed at Jameela]

Two main conceptions of play emerge from this discussion. The first portrays play as a formal, juridical process, to which players are subject (see Jameela's contributions). The second presents play as primarily affective, subjective, and driven by players' transitory desires rather than rules (see Michelle's and Helen's contributions in lines 42–55).

In her first contribution, Jameela emphasises the obligation involved in following rules: 'you have to . . . '. Her comments portray rules as objective, rational and impartial: sets of injunctions to be followed. All players are equal with respect to such rules (i.e., they apply equally to everyone). Consequent upon this view is that the hierarchical relations characteristic of nonplay classroom interaction no longer apply. Jameela tells Andrew what he *has* to do, speaking to him as a fellow player rather than a teacher. She does this on the basis of the role-playing 'game' which Andrew sets up at the start of the discussion (lines 3–5, 'If I was the Martian . . . ') as a way of persuading students to participate in academic analysis—it is noteworthy that in order to interrupt the game of Snakes & Ladders, another game is set up to retain students' cooperation. Jameela here then abides by Andrew's request to act playfully, according to the rules of the role-playing exercise he establishes.

This view of play as a set of rules to which individuals are (equally) subsumed is maintained by Jameela throughout. Her intervention in line 17 ('I think you can play with one person yeah?') suggests that her concern is with the consistency of rules rather than the pleasure of play(ers). Other students present a similar construction of play at the start of the discussion. Alisha, for example, focuses on the requirements of the game (line 15—'[It] takes several people'), with Helen also pointing to the contract binding players to organise their play as a fixed sequence (line 13—'you have to take turns to go around').

The portrayal of rules as conventions to which players subsume themselves collectively places emphasises on the power of those rules rather than the interests of players. According to Sutton-Smith (1997), this conception of play can be understood in terms of a rhetoric of power. Within a rhetoric of power, rules of play which are socially made appear objective and impartial. They serve to arbitrate competing interests rather than serve players' subjective interests or motivations.

A second conception of play emerges in counterpoint to the first, and is based upon the portrayal of rules as regularities of natural behaviour rather than universal laws. Although this representation of play emerges within a discussion directed by the adults, it serves to undermine their authority. This can be seen in the use of a different range of signifiers of play; rather than responding to the requirements of Andrew's role-playing scenario, play here interrupts the adults' endeavour to deconstruct Snakes & Ladders in an orderly manner—for example, through giggling, laughing, physical gestures and providing 'foolish' answers to serious questions.

This second conception of play is particularly noticeable from line 42, although it seems to be introduced by Michelle's comments a little earlier, which counter Andrew's efforts to obtain collective, clear understanding of the rules of Snakes & Ladders; Andrew's laughter in line 39 results from his attempt to pin Michelle down on what the rules are, as she disagrees with Andrew's reformulation of her own statement. Here, rules are not so much subsumed to as decided by the players *ad hoc*. This conception emerges more distinctly when Alisha, Michelle and Helen invent new rules which inverse the actual rules of Snakes & Ladders (e.g., significance is given to the colours of the snakes). These new rules are not intended to parallel the original, but to subvert the effort to examine games in terms of formal systems and situational obligations; the laughter in lines 46 and 52 emphasise that the alterations are precisely intended to be meaningless and arbitrary.

Like Andrew in lines 3–5, Alisha, Michelle and Helen here also create another game, but one precisely that refutes his role-playing scenario. They establish their own 'metagame' in counter-position to his. In effect, they are saying 'I can do meta too'.

This second conception of play can be understood in terms of Sutton-Smith's rhetoric of frivolity, in which play is applied as a term to describe the activities of the idle and foolish. Here of course, the students apply it to themselves to challenge the requirements placed upon them. Sutton-Smith argues that this rhetoric rebuts the seriousness of 'regular' play, by trivialising it; it is so frivolous it inverts the frivolity of play itself. One could argue that in rebutting the adults' efforts to take play seriously (note my refocusing questions lines 57 and 60 to get the discussion back on task), the students here satirise the very foolishness of the role-playing scenario and the adults' attempt to make meaning from play. Their banter about the possible significance of the snakes' colours could be described as a carnivalesque discursive strategy to inverse the processes and social relations by which a discussion of this kind makes sense. It also breaks down the opposition between subject and object implied by the role playing scenario; although the adults seek to have a discussion *about* play, this second construction of play makes the academic discussion the very subject of playful laughter and resists the endeavour to fix play within agreed rules.

The two conceptions of play emerge on the basis of conventionalised power relations between students and teachers. However, they also position the students differentially in relation to each other. In lines 61 and 62, Michelle counters my attempt to recuperate the discussion (line 57) by banging the counters on the table. When Jameela mimics her gesture (lines 66–67), she is undermined by Alisha who accuses her of cheating (line 69); having positioned herself as Andrew's ally, Jameela is marginalised by the other three students. Michelle's physical clowning plays to an appreciative audience, one from which Jameela is pointedly excluded.

In this section I have endeavoured to make two points. First, conceptions of play emerge within, or as, social and power relations. Jameela's and Michelle's competing conceptions position them differently with respect to the adults and the other students. Second, playing, analysing game play, and making games (not covered here) are tightly interlinked activities. Analysing play here becomes a form of playing. This suggests that what demarcates play from nonplay is not so much formalised characteristics as the state of social relations at any one time. Different interests in a group situation generate different boundaries for play. This raises an interesting question about whose interest predominates in studies of play, either in schoolwork or in academic writing.

CONCEPTIONS OF PLAY IN ANALYSING PRINCIPLES OF GAME DESIGN

Following the session described above, two more focused on Monopoly and Cluedo. Over those 3 weeks, the group gained fluency in talking about game components and structures, and became more experimental in editing games. However, discussions remained focused on the instances of games we had played, rather than ludic structures more generally. For example, we discussed the significance of various penalising and rewarding squares in Monopoly (the income tax square, the parking fine square), but did not move onto reviewing the distribution of values in game design generically. To facilitate this process of abstraction, I reviewed the points made with respect to each of the games we had played and organised them under general headings: the role of chance, the choices available to players, the ordering of space, the organisation of time, the way in which information is provided/withheld/obtained, the skills required by players, and the game's material components. Aspects of the games relating to each of the headings were summarised in a table; for example, I summarised how space was ordered and the basis on which it was moved through in each of the games. The entry for 'Time' read as follows:

Game	Time
Snakes & Ladders	Get a head start if throw a six/Take turns
Chess	Same colour always goes first/Take turns
Monopoly	First trial run round the board (first one gets first choice of property)/Take turns/Have the right to buy first/Miss a turn or more (in jail)/Get out of jail card
Cluedo	Same character always goes first/Take turns/ Time to get to a particular room/Guessing answer before someone else

The completed table was given to students at the start of the subsequent session. Pairs of students were assigned to each heading and asked to present the main points under it. The transcript below is a sample from the ensuing discussion. Jameela and Zawadi here are talking about the column headed 'Time', as shown above. My purpose in discussing this transcript is to highlight differences with the session on Snakes & Ladders with respect to conceptions of play.

1.	*Zawadi*	We think time is important in Snakes & Ladders,
2.		because if you throw a six you get a head start, which means that
3.		you . . . It depends on the chance. It depends on how
4.		many sixes you throw, but the more sixes you throw, the
5.		Sooner you'll get to the finish line. Because it's all in
6.		turn, if anyone else gets a six before you, going in turn, it will
7.		take you longer to start and therefore it will take you longer to get
8.		to the finishing obviously. Who gets to finish first is the
9.		winner. So Snakes and Ladders is mainly based on chance
10.		and time. And that's about it. But Chess, time, time . . .
11.		I don't think time is as big a deal in Chess, it's just
12.		that the same colour goes first and that's the only
13.		thing. And then Monopoly, same thing, you have to get a six
14.		before you start, because everyone is going in turns, it still
15.		may take some time, a long time for you to get a six.
16.		Yeah.
17.	*Jameela*	When you are in jail
18.	*Zawadi*	Yeah, when you are in jail, it wastes time because everyone else
19.		gets another go and they can roll the dice and get first and do
20.		more stuff while you are stuck in jail.
21.	*Caroline*	So what stuff do they do? What's the point of sending people to
22.		ail, why is that a punishment?

23.	*Alisha*	Because you have to pay to get out?
24.	*Jameela*	To delay them.
25.	*Zawadi*	Yeah, to delay you. To delay you while others can catch up or
26.		go further. Which makes it more exciting, it makes the
27.		game more exciting.
28.	*Andrew*	Why does it make the game more exciting, why does it not just
29.		make it more boring?
30.	*Zawadi*	Because it's . . .
31.	*Andrew*	If you are being delayed
32.	*Zawadi*	Because it's . . .
33.	*Jameela*	Not more exciting for you but more exciting for us.
34.	*Zawadi*	No it's exciting for you as well!
35.	*Andrew*	Because?
36.	*Zawadi*	Because [pause]
37.	*Alisha*	Because it's boring if you know where you are and you are never
38.		going to stop.
39.	*Zawadi*	Yeah, yeah, it gets boring if you know that, OK, I go round this
40.		twice and nothing happens to me, it gets boring. But if you know
41.		something is going to happen you try and get away and try to . . .
42.		You try to [pause] . . .
43.	*Jameela*	Avoid it.
44.	*Zawadi*	Avoid it, yeah, so it makes it more interesting. And when you get
45.		caught, it's a funny feeling.
46.	*Alisha*	It can get annoying but still . . .
47.	*Zawadi*	It can get annoying yeah.
48.	*Alisha*	But I think it's good to get people sometimes.
49.	*Zawadi*	Yeah it makes the game more interesting.
50.	*Alisha*	And it makes it fairer as well.
51.	*Andrew*	OK, so time is related to how you feel about the game, it's related
52.		to skill and trying to find things, and to chance. I mean some of
53.		these things . . .
54.	*Zawadi*	Yeah, specially the dice games. And Cluedo as well. I didn't play
55.		Cluedo at all.
56.	*Andrew*	What's time in Cluedo?
57.	*Zawadi*	Characters go first in Cluedo but I don't really know about that.
58.		And everyone takes it in turns and you have to . . . She
59.		can tell you about Cluedo
60.	*Andrew*	Time in Cluedo.
61.	*Jameela*	Time is very . . . I don't know, it's like strategic, because you
62.		are waiting for people, to move and stuff, and waiting for people
63.		to tick off what's on their sheet, and you don't know when someone is
64.		going to say 'ooh, I won, I know what's going on' so it's very annoying
65.		at some point to keep waiting. They are waiting till it's
66.		their turn to move.
67.	*Zawadi*	Isn't that the same? Everyone is waiting and you don't know

68.		who's got the answer.
69.	*Caroline*	Yeah.
70.	*Jameela*	Because it keeps happening/
71.	*Zawadi*	over and over/
72.	*Jameela*	over and over/
73.	*Zawadi*	Till you get bored, but it is interesting.
74.	*Andrew*	So it's about the balance of time. You know, up to a
75.		point you can make things more exciting but beyond that point it
76.		gets boring.
77.	*Zawadi*	Yeah.
78.	*Caroline*	It's quite suspenseful, isn't it, because last week you were saying that
79.		you had lots of ideas about who did it and with what, but it wasn't
80.		your turn, so everyone was going round and there was a sense of
81.		suspense.

In this passage, the rules of play are described in terms of a conjunction between the meaning they have in organizing play and the pleasures this affords players. The distribution and function of time as an abstract element is discussed in terms of rules, but also related to the quality of players' subjective experience. Whereas in the first session, views of play as rule-governed and as subjectively meaningful were polarized, here the two aspects seem closely related. In the exchange in lines 33–34, for example, Zawadi argues that even when rules work against a player, he or she continues to have an interest in upholding them ('it's exciting for you as well!'). Rules are therefore maintained by players to serve their own purposes. Conventions of play are upheld because of the meanings they have for players rather than juridically imposed. Such meanings are described in terms of affect (e.g., lines 44–45, 73—'you get bored', 'it's a funny feeling', 'it makes the game more exciting') in the establishment of social relationships with other players (e.g., lines 18–20—'when you are in jail, it wastes time because everyone else gets another go'). Play here is a way of creating a specific kind of social bond, one based on fairness and competition (e.g., line 50—'it makes it fairer as well'), and which is expressive of emotional and sensual states.

Much attention is given to how people feel when they are playing (play is 'boring', 'exciting', 'funny', 'annoying'). Sutton-Smith describes a concern with subjective experience in play in terms of a rhetoric of the self, with emphasis placed on personal fulfilment, self-possession and self-expression. One way of interpreting this is to say that students' frivolous, carnivalesque attitude, identified in the first session with respect particularly to Michelle's and Helen's contributions, has been recuperated and made to serve the educational ends of the research activities. Games and play have been brought under the aegis of the curriculum and lost their more subversive potentials in children's folklore. Walkerdine (1999) notes that one of the dangers of the notion of media literacy is that it 'tends to build upon pre-existing strategies and technologies which stress rationality as a counter to the irrationality of the dangerous classes' (p. 9). Sutton-Smith makes a related point in arguing that efforts to study play in schools 'takes it out of its original collective context' and is often 'a further effort at the upper status domestication of the depraved' (1982, p. 195).

The purpose of this fieldwork was to study games and game play. As Sutton-Smith notes, the study of play can romanticise and domesticate it. However, it can also develop capacities to use games and game play as a means of expression and communication. Students here have developed a vocabulary and analytic practice for understanding how games function as structures, productive of meaning and emotion. This counters another romantic notion, that children inherently master playful social practices, that such a capacity is an innate ability, an expression of inner emotional states. The transcript above, however, indicates that a particular kind of social and analytic practice can sustain potentially more complex and sophisticated understanding of games and play, which enable children to participate in game play as a significant social activity more fully.

It should also be emphasised that the rhetorics of power and frivolity did not disappear from the club; students continued aligning themselves with or against the adults, with and against each other. However, they did so on the basis of a more informed approach to game design and analysis. The study of play may domesticate it in the interests of social elites, as Sutton-Smith argues. However, in this section, my aim has been to suggest the ways in which it can also generate more elaborated understanding of games as an expressive form. As researchers, we saw this as important in creating conditions in which students could design games, and so become designers of play, as well as consumers of games.

CONCLUSIONS

In this chapter, I have endeavoured to make two main points. First, play is a particular way of producing meaning, with interpretation of what play 'is' subject to negotiation and the formation of social allegiances, and thus identities. I drew on the anthropological literature on play to suggest limitations in conceptions of play as an activity which is intrinsically motivating rather than meaningful in relation to the development of identity in hierarchical group relations. Second, although one cannot study play in a disinterested way, it is nevertheless possible to develop a more sophisticated vocabulary about its components and operations, notably in the case of games.

I have not commented here on the transition from board game analysis to computer game-making. My intention was not to provide a course template, but to suggest ways in which a vocabulary and analytic practice can be developed to think about games and game play more systematically. In subsequent fieldwork, the approach we took to studying games evolved, and adapted for use in classrooms. We also sought to reintroduce a concern with representation, as inseparably linked to ludic design. The transition to software-based production introduced new complexities and possibilities, relating to the semiotic, material resources involved in digital production. The process of analysing and discussing game design principles however maintained attention on the functioning of computer games as a genre rather than on the programming environment per se.

The negotiation of social allegiances and identities continued when students started working with digital production tools, but on the basis of different resources. In designing computer games, students established different kinds of social affiliations, positioning themselves for example as fans of *Star Wars* or as expert programmers, in counterdistinction to others, by incorporating content from fan-maintained websites or by devising fearsomely complex rule structures (Pelletier, in press). Designing games itself became a form of play, with students often making games to play *with* others, rather than for others to play (this was achieved by creating games which were extremely difficult to play, and in which progress could be achieved only with the help of the designer [Pelletier, 2007]). This was often a source of tension, or perhaps more accurately, a site of negotiation, with teachers, researchers and students working with different notions of what constitutes a game and game play, notions which evolved over time and across different institutional contexts, and in response to changing social strategies.

This perhaps points to some of the implications of digital technology for learning. Much of the literature on computer games in education has focused on the motivational benefits of computer games. I have focused here on the way in which games sustain practices of signification. The introduction of games, games play and game making into educational sites is significant not primarily in terms of new ways of delivering the curriculum, but in the forms of social interaction which genres such as digital games give rise to, and the social practices and social identities that emerge in tandem with such genres, in the classroom and beyond it.

NOTES

1. The research project 'Making Games' was a collaboration between researchers in media education at the Institute of Education, London, and Immersive Education, a company which develops educational production software. The project was led by Professor Buckingham and Dr. Burn, with myself as the project manager. It was funded by the Paccit Link programme.
2. This is available with the software, from Immersive Education.

REFERENCES

Bateson, G. (1976). A theory of play and fantasy. In J. S. Bruner (Ed.), *Play* (pp. 119–128). Harmondsworth, UK: Penguin. (Original work published 1955)

Bruner, J. S. (1976). *Play: Its role in development and evolution.* Harmondsworth, UK: Penguin.

Buckingham, D. (2003). *Media education: Literacy, learning and contemporary culture.* Cambridge, UK: Polity Press.

Buckingham, D., Grahame, J., & Sefton-Green, J. (1995). *Making media: Practical production in media education.* London, UK: The English and Media Centre.

Burn, A., & Durran, J. (2007). *Media literacy in schools: Practice, production and progression.* London: Paul Chapman Educational Publishing.

Caillois, R. (1967). *Les jeux et les hommes* [*Man, play and games*]. Paris: Gallimard.

Carr, D., Buckingham, D., Burn, A., & Schott, G. (2006). *Computer games: Text, narrative and play.* Cambridge, UK: Polity Press.

Dawes, L., & Dumbleton, T. (2001). *Computer games in education project.* Retrieved July 15, 2004, from http://www.becta.org.uk/page_documents/research/cge/report.pdf

Department for Education and Skills. (2003). *Towards a unified e-Learning strategy—Consultation Document.* Retrieved June 6, 2007, from www.dfes.gov.uk/consultations

Department for Education and Skills. (2005). *Computer and video games in curriculum-based education.* London: DfES.

Ehrmann, J. (1968). Homo Ludens revisited. In J. Ehrmann (Ed.), *Game, play, literature.* (pp. 31–57). New Haven, CT: Eastern Press.

Farley, R. (2006). *Playing explorers: The re-enactment of legendary sea voyages.* Unpublished doctoral dissertation, University of Wales, Cardiff.

Freud, S. (2003). *Beyond the Pleasure Principle, and other writings* (new ed.). Harmondsworth, UK: Penguin Books.

Gee, J. (2003). *What videogames have to teach us about learning and literacy.* New York: Palgrave Macmillan.

Gee, J. (2004). *Situated language and learning: A critique of traditional schooling.* London: Routledge.

Groos, K. (2005). *The play of animals.* (E. L. Baldwin, Trans.). Whitefish, MT: Kessinger Publishing Co.

Hobbs, R. (1998). The seven great debates in the media literacy movement. *Journal of Communication, 48*(1), 16–32.

Huizinga, J. (1970). *Homo ludens: A study of the play element in culture.* London: Paladin.

Kirriemuir, J., & McFarlane, A. (2004). *Literature review in games and learning.* Bristol, UK: Nesta Futuralab.

Pelletier, C. (2007). *Learning through design: Subjectivity and meaning in young people's computer game production work.* Doctoral dissertation, Institute of Education, University of London.

Pelletier, C. (in press). Producing difference in studying and making computer games: How students construct games as gendered in order to construct themselves as gendered. In Y. Kafai, C. Heeter, J. Denner, & J. Sun (Eds.), *Beyond Barbie and Mortal Kombat: New perspectives on gender, games and computing.* Cambridge, MA: MIT Press.

Piaget, J. (1976). The rules of the game of marbles. In J. S. Bruner (Ed.), *Play* (pp. 413–441). Harmondsworth, UK: Penguin. (Original work published 1965)

Salen, K., & Zimmerman, E. (2003). *Rules of play: The fundamentals of game design.* Cambridge, MA: MIT Press.

Sutton-Smith, B. (1982). Play theory of the rich and for the poor. In P. Gilmore & A. A. Glatthorn (Eds.), *Children in and out of school* (pp. 187–205). Washington DC: Harcourt Brace Jovanovich and Center for Applied Linguistics, University of Pennsylvania.

Sutton-Smith, B. (1984). Text and context in imaginative play and the social sciences. In F. Kessel & A. Goncu (Eds.), *Analysing children's play dialogues.* (pp. 53–70). San Francisco: Jossey Bass.

Sutton-Smith, B. (1997). *The ambiguity of play.* Cambridge, MA: Harvard University Press.

Walkerdine, V. (1999). Violent boys and precocious girls: Regulating childhood at the end of the millenium. *Contemporary Issues in Early Childhood, 1*(1), 3–23.

Whiteman, N. (2007). *The establishment, maintenance and destabilisation of fandom: A study of two online communities and an exploration of issues pertaining to internet research.* Unpublished doctoral dissertation, Institute of Education, University of London.

11 Digital Cultures, Play, Creativity
Trapped Underground.jpg

Victoria Carrington

INTRODUCTION

Four bombs exploded around central London on the morning of the July 7, 2005. The bombs killed 52 commuters travelling on the public transport network and injured more than 700. On that day, the BBC received 20,000 emails; 3,000 photographs; 1,000 video and still images; and 3,000 text messages related to the events of the morning. A mobile phone camera captured what became the iconic image of the day. Adam Stacey and Keith Tagg were on London's Northern Line travelling past Kings Cross Station when they were caught up in the explosion aftermath. As the carriage evacuated via the tunnel system, Keith took a picture of Adam on a camera phone (Dear, 2006). Highly emotive and timely images take their place in history—the student standing in front of a tank in Tiananmen Square in 1989; Nick Ut's 1972 Vietnam War photo of then 9-year-old Kim Phuc running naked and badly burnt by napalm; heavy smoke billowing from the Twin Towers in New York; and now, the trapped underground.jpg. This image of being 'trapped underground' was uploaded onto a moblog (moblog.co.uk) that morning; since posting it has been viewed 150,748 times on that site alone (as of May 2007) and has become the image most associated with that fateful morning.

What is it about this particular image that captured the public imagination and why is this important in relation to our understandings of literacy and literacy education? To address these questions I plan to 'play' with the work of Theo van Leeuwen (2001) in drawing selectively from iconography and Barthian semiotics to build a layered analysis of this particular digital image. According to Van Leeuwen (2001):

> These two approaches ask the same two fundamental questions: the question of representation (what do images represent and how?) and the question of the 'hidden meanings' of images (what ideas and values do the people, places and things represented in images stand for? (p. 92)

He goes on to suggest that while semiotics looks at an image and examines what is depicted within it, iconography 'pays attention to the context in

Figure 11.1   Trapped underground

which the image is produced and circulated, and how and why cultural meanings and their visual expressions come about historically' (Van Leeuwen, 2001, p. 92). Therefore a critical reading of an image has the potential to unlock social and cultural values, to understand what Barthes (1981) called the 'punctum' or the unexpected power of a photograph to take on meaning beyond itself, to become iconic.

In this particular chapter, I am interested in both how and why this particular image has become key. The combination of iconography and semiotics outlined by van Leeuwen was developed around planned, posed images ripe with intentional and unintentional meaning. In this instance, I am interested most particularly in the networks of context—social, political, technological—that led to the iconic status of this photograph. Following this, I plan to examine this image in the context of an emerging notion of participatory culture with its emphasis on collaborative learning, play and appropriation and finally, consider how these issues are relevant to those of us interested in literacy and literacy education.

IMAGES IN CONTEXT: ICONOGRAPHY

The London bombings took place against the backdrop of increasing global tension in the wake of Bush-era U.S. foreign policy and growing attention to a changing religious and cultural profile (Leiken, 2005) in many parts of the world. At the same time, key publications (Huntington, 1997) and discourses pointed to a growing tension between fundamental belief systems. The events of September 11, 2001 had served to magnify these concerns and in the intervening years the western world experienced a heightened level of fear and alert for terrorist attacks. Mass media and political mileage has been made from deploying discourses designed to sharpen everyday perceptions of tensions, fear, instability and awareness of the 'other'. Reinforcing a strident good versus evil dichotomy, U.S. President George W. Bush outlined the threat from 'our terrorist enemies':

> They seek to impose Taliban-like rule country by country across the greater Middle East. They seek the total control of every person, and mind, and soul, a harsh society in which women are voiceless and brutalized. They seek bases of operation to train more killers and export more violence. They commit dramatic acts of murder to shock, frighten and demoralize civilised nations, hoping we will retreat from the world and give them free reign. They seek weapons of mass destruction to impose their will through blackmail and catastrophic attacks. (Retrieved October 27, 2004, from http://whitehouse.gov/news/releases/2004)

In 2006, the President noted in relation to the activities of the Homeland Security Department:

> 'We are still a nation at risk. Part of our strategy, of course, is to stay on the offense against terrorists who would do us harm' (Retrieved February 8, 2006, from http://www.whitehouse.gov/infocus/homeland/).

The US National Defense website currently proclaims:

> The Best Way To Protect Our People Is To Take The Fight To The Enemy. In Afghanistan, a regime that gave sanctuary and support to al Qaeda as they planned the 9/11 attacks has come to an end. And in Iraq, we removed a cruel dictator who harbored terrorists, paid the families of Palestinian suicide bombers, invaded his neighbours, defied the UN Security Council, and pursued and used weapons of mass destruction. (Retrieved June 20, 2007, from http://www.whitehouse.gov/infocus/defense/)

This global threat was reflected in speeches made by the then Australian Prime Minister, John Howard, who argued in 2003 that:

> As more rogue states acquire chemical and biological weapons, so the danger of those weapons falling into the hands of terrorists will multiply. If terrorists acquire weapons of that kind, that would represent a clear, undeniable and lethal threat to a western nation such

as Australia. . . . We do live in a different world now, a world made more menacing in a quite frightening way by terrorism in a borderless world. (Retrieved March 10, 2005, from http://www.ausint.com/au/subs/view_top_pm_announcement_030318.htm)

Contemporary culture is saturated with risk. While the United States and many of its allies perceive themselves to be at risk, many groups and individuals in other parts of the world feel strongly that they are at risk from the United States. Within this process, contemporary westernized notions of risk have been constructed as discourses of self-regulation and increased government power. This tension has manifested in the everyday in, for example, changes to airline security, the erosion of civil rights across many contemporary nations, tensions around Afghanistan, Iran and North Korea, and the establishment of the infamous U.S. Department of Homeland Security as well as a United Nations Commission on Human Security (2001, http://www.humansecurity-chs.org). In people's streets and homes, risk discourses manifest for example, around topics ranging from global warming and water shortages to fear of young men wearing hoodies; from concern with the long-term effects of food additives to the link between reduction in play areas for children and childhood obesity. While refracted through the particularity of the local, risk discourse is ubiquitous. It is in this climate that the London bombings took place and that this particular image was taken. Almost paradoxically, the image also locates itself in the everydayness and diversity of commuter life in London and other large urban centres. Every weekday, London's northern tube line carries over 660,000 passengers (over 206.7 million passengers each year; Transport for London http://www.tfl.gov.uk). There is a very good chance that many of the people who viewed this image have themselves travelled on London's underground system or one very much like it.

The 'everydayness' of the moment was reinforced by the use of mobile phone technology to capture and distribute the image. Mobile phone ownership in the United Kingdom is rife—it is estimated that there are over 62.5 million phones in the United Kingdom for a population of 60 million (Collins, 2006). This technology has become thoroughly embedded in the everyday and has impacted on the everyday social and often economic practices of individuals and communities (see, e.g., Cronin, 2005; Ito, Okabe and Matsuda, 2005). The mobile phone on which the image was captured was a taken for granted, everyday item to be found in the hands, pockets or handbags of the majority of commuters. Once taken, the image was forwarded to a well-known blogger. From there it was emailed and forwarded between blogs and individuals and was then reproduced in the mainstream media. The speed with which this became an early lens into an appalling tragedy was marked.

As a digital image, the photograph could be rapidly disseminated. Bridging a range of media, the .jpg started appearing on other blogs within the hour, mainstream media by the lunchtime news, was undoubtedly emailed from one person to the next, linked to numerous blogs, and was subsequently voted one of the *Time* magazine's Best Photos of the Year for 2005. It was from this event and the way in which the public used the portable digital technologies in their pockets to transmit information to each other, to authorities and to mainstream news media, that attention was drawn to a new role for the digitally connected citizen. Emily Bell, editor-in-chief of *The Guardian* newspaper (2006), notes that:

> The London bombings were the first domestic news story where the most significant coverage came from people at the scene—via mobile phones—rather than from established news outlets. It is striking that on September 11, 2001, the mobile as a means of instant communication played a hugely significant part in piecing together the day's events, yet there are no images from inside the twin towers as the picture phone was not habitually used or available. (Retrieved February 26, 2007, from http://www.guardian.co.uk/commentisfree/story/0,,1815710,00.html)

The image works to connect the local and global in a particular moment. The affordances of digital technology meant that this moment could be shared unendingly with vast numbers of people in local and global sites within a few other moments.

The title *trapped* underground plays a role in connecting this particular image to a broader set of meanings. It could have been named something else, for example, 'leaving the northern line' or 'standing near a doorway' or 'I hate Wednesdays'. It wasn't. The words 'trapped' and 'underground' are associated with an entire genealogy of other narratives and media reports both fiction and nonfiction. From children's adventure books to television biopics, the emotive power of the label trapped underground is high. The photo also reinforces a deeply embedded national narrative about British stoicism (see, e.g., Epstein, 2005) and resolve in the face of adversity. The seeming lack of panic, the orderly exodus from the underground tunnel, the photographer taking a moment to take a photograph, feeds upon and into this narrative. What was interpreted as quiet defiance in the face of terrorist attack was a point of national pride. Adam Stacey's interview account of the incident describes a range of reactions from calm to panic (Dear, 2006); however, the image captured on the mobile phone does not indicate extreme panic amongst passengers. Britain's Prime Minister, Tony Blair reflected this sentiment:

> When they try to intimidate us, we will not be intimidated. When they seek to change our country or our way of life by these methods, we will not be changed. When they try to divide our people or weaken our resolve we will not be divided and our resolve will hold firm. The

purpose of terrorism is just that—it is to terrorise people. And we will not be terrorised. (Epstein, 2005, Retrieved February 28, 2007, from http://www.abc.net.au/am/content/2005/s1409785.htm)

In this, the photographs sit squarely within a particular ideology and cultural self-narrative. This is not unusual. Sturken and Cartwright (2001, p. 13) observe 'over time, images have been used to represent, make meaning of, and convey various sentiments about nature, society and culture as well as to represent imaginary worlds and abstract concepts'. Thus, while the trapped underground.jpg captures a moment, it does so within a network of meaning and technology. Adam Stacey, via his representation in this particular image, takes on a Barthian mythic status as he connotes the everyday commuter, British stoicism under extreme pressure and the innocent bystander of the modern malaise of terrorism.

THE IMAGE ITSELF: SEMIOTICS

Photographs are more than the supposed 'capture' of an objective fact. Barthes (1981) argued convincingly that they are 'complexly coded cultural artefacts' (Van Leeuwen & Jewitt, 2001, p. 89). The ways in which the trapped underground image constructs meaning and the ways in which it is interpreted are central to this discussion. As described earlier, the photograph is an artefact produced within a particular technological and cultural landscape and this positioning influences framing, composition, subject and the mode of capture. In turn, these features allow us to better understand the image and the meanings we take from it. This particular image was created using a mobile phone camera. Like all technologies it carries with it a set of capacities and limitations. While undoubtedly useful in capturing images on the run so to speak, the cameras in the current generation of mobile phones generally have poor resolution, are sensitive to poor light conditions and have a time lag between pressing the 'shoot' button and image capture. Combined, these features increase the likelihood of blurry, grainy images and require that subjects are positioned relatively close to the lens. Keeping all this in mind, the semiotics of this image remain powerful. Adam looks directly out from the picture. His eyes make contact with our eyes. His eyes squint into the camera leading us to imagine that the graininess of the image is matched by smoke and dust in the tunnel. Blurred commuters are visible in the background, some still inside the carriage waiting to leave. Standing directly between the viewer and the drama playing out behind him, Adam demands our attention to him and to the situation in which he and his fellow commuters find themselves. As Kress and Van Leeuwen (1996, p. 122) note, the direct gaze 'creates a visual form of direct address. It acknowledges the viewer explicitly'. The vector of his hand leads directly to his face and eyes, reinforcing the relationship between viewer and image and the authenticity of the image and what it captures. Adam's is the only face entirely visible and facing the camera in the style of a self-portrait, documenting his location at that moment in time, the decisive moment (Cartier-Bresson, 1952) when the significance of the event and the organization of forms intersect. In part, the choice of proximity is determined by the technology used to take the image. The dark textured look is amplified by the contrast with the internal light of the tube train door against which Adam is silhouetted. He is geographically, temporally and culturally located by this particular image. Two men passing directly behind Adam are wearing suits, their jackets and white shirt collars visible in the gloom. Men in suits carry a particular cultural significance—commuters on their way to professional jobs in the city, people with responsibility and respect. The disruption seems all the more confronting against a backdrop of white-collar commuters.

This particular portrait, grainy, framed against the backdrop of a stranded and disabled tube train creates a direct connection between Adam in situ (and based on our shared knowledge of the tube system and terrorist attacks, clearly at risk) and those of us who view the picture from the safety of our personal devices or printed document. The dark graininess of the background, an effect of the dust, lack of light and the camera's characteristics, may even serve to enhance the feeling of intimacy and connection (Lister & Wells, 2001) as well as to suggest authenticity. The use of mobile phone technology and the status afforded the images created and distributed via its affordances parallels in some ways the authenticity claims made on behalf of video in the early years of its use. Early documentary work and independent film were closely associated with relatively low-specification, consumer quality video equipment and footage where 'directors used grainy black-and-white film, hand-held cameras, and long takes to capture unscripted action as it unfolded spontaneously in "real situations"'; the reality television shows of the 1990s used hand-held and surveillance cameras to depict 'reality' (Sturken & Cartwright, 2001, p. 287). In 1999, the movie *The Blair Witch Project* used this link between authenticity and video documentary-style footage as the hook for what become a cult horror movie with a tagline that announced: 'In October of 1994 three student filmmakers disappeared in the woods near Burkittsville, Maryland, while shooting a documentary. . . . A year later their footage was found' (Retrieved February 28, 2007, from http://www.imdb.com/title/tt0185937/taglines). The unprocessed, raw quality of this work lent it an automatic authenticity and sense of truth. As Van Leeuwen and Jewitt (2001) note 'whatever an image depicts or show us, the material means and medium employed to do so have a bearing upon which qualities of the depicted thing or event are foregrounded' (p. 89). The trapped underground.jpg recaptures this link between production quality, the everyday nature of the technology, and truthfulness.

PARTICIPATORY CULTURE

Recognizing a cultural shift to the visual and spatial, Kress and Van Leeu-wen, separately and together, have argued convincingly that we are now a semiotic rather than book-bound (Kress & Van Leeuwen, 1996, 2001; Kress, 2003) culture. Scollon and Scollon (2003) and Barton and Hamilton (2003) argue with equal conviction that literacy practices are localized and spatialised. Trapped underground.jpg is an example of the move to the visual in our culture as well as the localized, spatialised and technologised nature of literate practices. The image also captures a key feature of glocalisation—where global and local intersect dialectically. For those of us interested in literacy, this is an important insight. Literacy practices have always pivoted on the local and on the available technology (even if that technology was a graphite pencil) just as they have been responsive to larger cultural and social issues. Changes in available technologies do not alter this, but they do afford potentially new practices with text that are important to note.

Being competent and able to participate in one's community requires a cocktail of skills around technologies and the production, dissemination and 'reading' of innumerable types of text. These skills and technologies are foregrounded in popular culture as well as in the mainstream broadcast news media and are reflected in art. The rapid rise of trapped underground. jpg to worldwide recognition and the emergence of other sites of participa-tion that require new interactions with changing forms of text are markers of this shift. The valued practices and capitals in many dominant social fields—corporations, education, government, military, popular culture—are shifting to include those linked to the types of skills and knowledges characteristic of digital texts and the technologies and practices that pro-duce them. It is increasingly valuable to be able to create multimodal texts that can operate across a range of platforms, to rapidly critique informa-tion from a range of sources, to move back and forward between basic skill in print literacies and skill in multiliteracies, to work in peer learning contexts and informal settings. Many of these skills and attitudes parallel those identified as core to engaging meaningfully in the civic; many may be performed communally. They are skills and practices that take their shape and meaning within social networks. Jenkins, Clinton, Purushotma, Robi-son, and Weigel (2006) outline these skills on Table 11.1.

While the communality of many of these practices is central to this view, it is also the philosophy underlying contemporary views of literacy as a socially situated practice. Of course, while Jenkins et al. (2006) suggest that students must be able to read and write before they can engage in the emerging participatory culture, many of us working in literacy educa-tion would now argue that this same participatory culture should form the context in which practices around text (including baseline reading and writing around print) are developed. There is not a neatly sequential and linear relationship between basic print skills and the analytic skills and

Table 11.1   Key Skills and Practices of Participatory Culture

Play	Capacity to experiment with one's surroundings as a form of problem-solving
Performance	Ability to adopt alternative identities for the purpose of improvisation and discovery
Simulation	Ability to interpret and construct dynamic models of real-world processes
Appropriation	Ability to meaningfully sample and remix media content
Multitasking	Ability to scan one's environment and shift focus as needed to salient details
Distributed Cognition	Ability to interact meaningfully with tools that expand mental capacities
Collective Intelligence	Ability to pool knowledge and compare notes with others towards a common goal
Judgement	Ability to evaluate the reliability and credibility of different information sources
Transmedia Navigation	Ability to follow the flow of stories and information across multiple modalities
Networking	Ability to search for, synthesize, and disseminate information
Negotiation	Ability to travel across diverse communities, discerning and respecting multiple perspectives, and grasping and following alternative norms

(Table 1—Jenkins et al. 2006, p. 4, reproduced with permission)

higher order interactions necessary for engagement with digital texts and the participatory cultures they allow (see Freebody & Luke, 1990) for a discussion of the four resources model). Indeed, if these basics are to have any meaning, they need to be developed in concert with the entire range of participatory practices and technologies in common use in our culture. Use of digital technologies and the access to enhanced participation they allow should not be understood as a 'higher order' activity. Rather, engagement with digital texts and technologies as well as older forms of text within a participatory culture of the type outlined by Jenkins should be a core aspect of becoming literate in contemporary culture. Importantly, Jenkins and his colleagues suggest that these activities are forming a new 'hidden curriculum' that will act to determine which students succeed and which don't. The ways in which we conceptualize literacy and provide opportunities to develop literate skills and knowledge must increasingly take account of the pedagogic characteristics of a participatory culture.

IMPLICATIONS FOR LITERACY

While the trapped underground.jpg does not itself fall within the category of an older alphabetic literacy—it is after all, a digitally captured image—it is nevertheless representative of important new ways of participating in our various communities and the ways in which print and image are increasingly woven together to make a powerful message; for instance, to send an image taken on a mobile phone camera requires an engagement with in-phone menus and instructions. The ability to create and disseminate a meaningful message within moments is a social as well as technical watershed. It also points to potential shifts in the valuations applied to particular ways of participating in contemporary culture. For literacy educators, shifting valuations around particular types of practices with a range of texts and technologies are important signposts. It is our responsibility to read these signs and understand their implications for our work with literacy.

SOME SIGNPOSTS

Aidan Hatch is 6. With the help and supervision of his parents, he creates a blog called Information About Stuff (http://aidanhatch.blogspot.com/). His profile notes:

> This is my very first website. I'm going to put some information on it. I REALLY LOVE COMMENTS. PLEASE SAY HI BEFORE YOU LEAVE!

The awareness of social interaction with an online community is clear, as is his knowledge of the mobility ('before you leave') that characterizes online navigation. The use of the word 'information' in the title is telling. Aidan's interest is on researching and finding out 'about stuff'. He shares this new information about topics of interest with his online audience. His ability to successfully compile and convey useful information in a way that is interesting and effective is reinforced by the comments he receives from an online audience. Audience members validate Aidan's skills at research and blogging and reinforce this by suggesting possible research topics for Aidan to pursue on his blog:

> Aidan your blog is fantastic! That Wobbegong shark looks pretty creepy. I never knew they existed. You seem to know alot about animals. I wonder if you could make a post about Meerkats? I saw Meerkat Manor on Animal Planet and would like to know more about them. (erin 8:44 a.m.)

Aidan's parents and the online community with which he interacts are scaffolding him into mastery of a range of skills and knowledges in relation to online texts, printed text and the ways in which the two are entwined to make meanings that are relevant. His blog incorporates text, some dictated to a parent, some typed in by Aidan as well as photographs which appear to be chosen by Aidan to match his topic with some of them found on other online sites. The use of a photo from Flickr is explicitly acknowledged and the photographer thanked. Aidan is being mentored into appropriate etiquette. Some of his spelling is nonstandard but he manages to convey his meaning; he is experimenting with the use of coloured font and underscoring. The blog also contains a podcast of Aidan and his father talking about the coming school year. Aidan is appropriating, mixing and matching a range of new media, with assistance and within a community, to undertake authentic tasks around creating meaning. This is powerful. 'Information about stuff' is being used to promote learning, to provide opportunities for Aidan and his family to engage with the technology, talk about projects and ideas, spend enjoyable time together. Clearly there is a not insignificant commitment of family resources to these activities. This commitment is a reflection of the value placed on learning how to use digital technologies and the texts and practices associated with them to participate effectively and ethically in a range of communities and quite possibly, a tacit recognition of the hidden curriculum suggested by Jenkins.

Aidan's younger brother Clay also has a blog, 'Clay's Special Website' (http://clayspace.blogspot.com/), also supervised and scaffolded by his parents. Demonstrating that they understand the issues of identity rolled up in any textual practice along with the specific characteristics of blogging, Clay's parents support Clay to build and maintain his own blog depicting his particular interests and knowledge. His experimentations with coloured font are highly visible, along with his interest in the collection and display of facts—facts about helmets, knights, castles and sharks.

Commenting on the phenomenon of kids and blogging on the well-known blog lifehacker.org, Brogan (2006) makes a series of comments about Aidan and his blog:

- He's 6.
- He likes changing the colours of the words every bit as much as the research.
- He must be using Google to research, or Wikipedia.
- He's using blogging tools.
- He's observing social software (his love of the comments).
- He's getting good follow-on feedback from readers, which drives him to research more.
- He's actually planning posts (Retrieved June 25, 2007, from lifehack. org/articles/lifehack/blogging-for-kids.html).

While Brogan makes some insightful comments, he is also signposting a route to Aidan's blog. Being discussed on another blog is an important indicator of esteem in the world of blogging. Readers of one site will click on the link to follow the trail of reference to the next. In effect, reading this site led me to Aidan's blog and Aidan's led me to Clay's. Aidan's blog also connects to the website of an author/friend of the family and other sites and people of interest to him. Aidan's blog is becoming part of an intricate and dynamic network of online sites with readers following the various hyper-link trails from one to the next.

Of course children are not limited to blogging. Young children are also making use of digital cameras—either the ones lodged in their mobile phones or dedicated cameras to capture images. These images are being posted by the children with the assistance (or not) of adults on blogs and dedicated sharing sites. The photo sharing website Flickr has a group called Little photographers—photos taken by children (http://www. flickr.com/groups/83295436@N00/pool/page6/). At time of writing there are 88 members and almost 400 photos listed in the group. One of these members is TripleH. He is 11 years old and is already taking part in the social networking that takes place around Flickr (Davies, 2006). While the range of images he has posted is wide, one of the set displayed on the Flickr site depicts a narrative told in sequence using text and image. Using posed Lego® figures, photographed in sequence, TripleH has constructed an adventure story which captures the trials and tribulations of a small band of intrepid Lego explorers as they attempt to escape a Lego-eating Bionicle®. The brave little toys are attached, losing an unfortunate victim to the jaws of the monster. They run, climb up the perilous side of a laundry basket to what they hope is safety only to be attacked again, lose another member of their team to their insatiable adversary and then climb back down the basket in an attempt to escape. TripleH uses the note-making feature (which allows annotations to be added in text boxes and overlaid onto posted photos) to build additional layering into the meaning of the photo and the narrative being constructed as well as to play with the software. This feature is not automatically displayed. In order to see the additional notes the viewer must hold the computer mouse over the image to make the note visible. This adds another layer of playfulness and potential for creativity to the text, allowing humorous asides by the characters depicted in the various frames of the narrative sequence. He also uses the notes to hyperlink to other sites and photographs. This Flickr sequence is a complex text drawing together a range of skills and textual practices and knowledge. Importantly, like the young bloggers described above, TripleH is engaging with a larger community who take his contributions seriously and give thoughtful feedback. One of the set of photographs, titled 'Uh oh' (capturing the moment when the brave Lego® men discover that the basket in which they sought sanctuary is not safe because their enemy can fly), was awarded the Barnfields Photography Club 'My Shot of the Week'

Table 11.2   Key Textual Skills and Practices for Participatory Culture

Capacity	Explanation	Aidan, Clay, TripleH
Play	Capacity to experiment with one's surroundings as a form of problem-solving	Playing with font, text, mixing and matching formats, trialling new media; playing with narrative structure using photographs and hidden notes.
Performance	Ability to adopt alternative identities for the purpose of improvisation and discovery	Aidan, Clay and TripleH are all foregrounding particular identities in these online sites—blogger, researcher, story maker, photographer.
Simulation	Ability to interpret and construct dynamic models of real-world processes	TripleH's narratives demonstrates understanding of the nature of linear time sequence; Clay and Aidan are demonstrating capacity for research process and interaction in multiple communities.
Appropriation	Ability to meaningfully sample and remix media content	Use of photos from other sites, combined with information and narrative to create new texts; use of range of aspects of digital technologies to develop and distribute texts
Multitasking	Ability to scan one's environment and shift focus as needed to salient details	Construction of narrative sequence by TripleH demonstrates focus on salient details; use of Flickr site demonstrates understanding of social networking site; Clay and Aidan are demonstrating their understanding of the ways in which blog entries are constructed and how to plan and construct salient posts.
Distributed Cognition	Ability to interact meaningfully with tools that expand mental capacities	Clay, Aidan and TripleH demonstrate an ability to interact with the internet and with other technologies (e.g. digital cameras). These expand their capacity to interact with a range of audiences and to develop a range of skills and capacities as they draw from a communal pool of knowledge.
Collective Intelligence	Ability to pool knowledge and compare notes with others towards a common goal	Use of social networking sites and engagement with communities

(continued)

Table 11.2   (continued)

Capacity	Explanation	Aidan, Clay, TripleH
Judgement	Ability to evaluate the reliability and credibility of different information sources	Research skills around areas of interest, using online and offline texts and other sources
Transmedia Navigation	Ability to follow the flow of stories and information across multiple modalities	Use of podcasting, images, text, hyperlinks
Networking	Ability to search for, synthesize, and disseminate information	Use of networks of distribute information gained by research (Aidan, Clay) and constructed narratives (TripleH)
Negotiation	Ability to travel across diverse communities, discerning and respecting multiple perspectives, and grasping and following alternative norms	Clay, Aidan and TripleH are being scaffolded and mentored into a range of communities. The feedback they receive from audience as well as direct mentoring is providing opportunities for developing skills to negotiate multiple communities.

(Table 2—adapted from Jenkins et al. 2006)

award, given in recognition of children's photography. On another site, the 7-year-old photographer is congratulated for the composition of the image of her sister; on another, the young photographer is congratulated for depth of field and colour.

Aidan, Clay and TripleH are learning, via mentoring within a social network and play, to experiment, undertake research, to plan, to attend to details like design, creativity and meaning. They are learning about collaboration and audience. Importantly, they are also learning that their activity is authentic and has meaning for others as well as for themselves and their immediate friends and families. These experience and skills are linked to the ways in which Aidan, Clay and TripleH construct their identities as learners and members of a family and various communities. The technologies and the social networks are directly connected to the development of particular kinds of literate skills and attitudes. Their parents are apprenticing them into a form of literate practice that is increasingly valued and rewarded. And importantly, many of the skills and dispositions Aidan and Clay are developing are transferable to other domains. There is no disconnect between offline and online literacies. Rather than mastering basic skills around print text and then moving to the online, these children are learning about text across off and online textual environments (see Cross, in this volume). Returning to the arguments outlined by Jenkins et al. (2006), these are the experiences and skills these children will require to be successful within an expanding participatory culture.

Table 11.2 shows how the experiences that Aidan, Clay and TripleH are accumulating map to the capacities identified as prerequisites for engaging in participatory culture by Jenkins et al. (2006). Reflecting the priorities of an increasingly participatory culture, the practices associated with these texts are clearly powerful and are becoming, if we follow Jenkins's argument, prerequisites for effective participation in contemporary culture.

IN CONCLUSION

I am committed to the notion that being literate is about being able to access and manipulate the texts created within one's society and to use them to meet individual and community needs and desires. Logically, as these texts and the technologies with which they are associated shift, so do the skills, knowledge and cultural knowledge prerequisites. It follows that we need to ensure that the children in our classrooms are given access to all the kinds of textual practice, around all the kind of technologies that are taking up powerful positions in our culture. Sometimes these practices will need to be around traditional forms of print; at other times they will be multimodal and/or digital.

I make no claim for the universality of these changes. There are individuals and communities where, for a variety of reasons, practices around text production and the ways in which they are valued have not shifted. There are, additionally, locally specific uptakes of any textual practice (Barton & Hamilton, 2003; Koutsogiannis, 2007). I do, however, believe that for many of the children growing up in contemporary English speaking cultures, and in many others where digital technologies are increasingly embedded, that practices and aspirations around textual production are shifting. At the same time, there is no suggestion here that print based literacy practices will erode away. Rather there is a need to increase the range of practices and understandings that attach to the notion of literacy. The trapped underground.jpg shows how culturally central the texts created by digital technologies have become in mainstream culture. This makes them important for those of us who research and think about literacy, not just those who work in the area of communications and media studies. The texts produced by Aidan, Clay and TripleH show how the practices associated with digital technology and participatory culture are making their way into family life. They are also signposting the types of text that are socially valued in many communities. Practices around socially valued text forms are core to being literate. These multilayered, digital texts and the textual practices associated with them are becoming a new addition to the set of powerful literacies. The digital literacy skills used to create, disseminate and discuss that particular multimodal text are the same ones that young people need in a world where global, local, national boundaries are increasingly contingent. These skills are the ones that will allow them to be creative, compliant, playful and transgressive, and to take their place in the participatory culture that Jenkins describes—a culture that will operate on and offline and all the spaces and places in between.

ACKNOWLEDGEMENTS

I wish to thank Jennifer Rowsell and Muriel Robinson for their insightful comments on earlier drafts of this paper.

REFERENCES

Barthes, R. (1981). *Camera lucida: Reflections on photography*. New York: Hill and Wang.

Barton, D., & Hamilton, M. (2003). *Local literacies: Reading and writing in one community*. London: Routledge.

Brogan, C. (2006). *Blogging for kids*. Retrieved June 25, 2007, from http://www.lifehack.org/articles/lifehack/blogging-for-kids.html

Bush, G. W. (2004). Speech to ArmyWar College, Carlyle, PA. Retrieved October 27, 2004, from http://www.whitehouse.gov/news/releases/2004

Bush, G. W. (2006, February 8). President Bush discusses Department of Homeland Security priorities. Retrieved June 26, 2007, from http://www.whitehouse.gov/infocus/homeland/index.html

Cartier-Bresson, H. (1952). *The decisive moment*. New York: Simon & Schuster.

Collins, B. (2006, May 21). Mobile madness consumes UK. *The Sunday Times*. Retrieved May 23, 2007, from http://technology.timesonline.co.uk/tol/news/tech_and_web/article722629.ece

Cronin, J. (2005, January 24). Africa's mobile entrepreneurs. *BBC News Online*. Retrieved May 23, 2007, from http://news.bbc.co.uk/2/hi/business/4145435.stm

Davies, J. (2006). Affinities and beyond: Developing ways of seeing in online spaces. *E-Learning*, 3(2), 271–234.

Dear, P. (2006). Images of 7 July: Tunnel horror. *BBC News Online*, 2nd July 2006. Available: http://news.bbc.co.uk/2/hi/uk_news/5102860.stm. Accessed 2nd May 2007.

Epstein, R. (2005). Blair pays tribute to British stoicism. Australian Broadcasting Commission, AM Radio, July 8, 2005, 08:08:00. Retrieved February 28, 2007, from http://www.abc.net.au/am/content/2005/s1409785.htm

Freebody, P., & Luke, A. (1990). Literacies programs: Debates and demands in cultural context. *Prospect: Australian Journal of TESOL*, 5(7), 7–16.

Howard, J. (2003). Transcript of the Prime Minister, The Hon. John Howard MP. Press Conference, Parliament House, Canberra, March 18, 2004. Retrieved March 10, 2005, from http://www.ausint.com/au/subs/view_top_pm_announcement_030318.htm

Huntington, S. (1997). *The clash of civilizations and the remaking of world order*. London: Simon & Schuster.

Ito, M., Okabe, D., & Matsuda, M. (Eds). (2005). *Personal, portable, pedestrian: Mobile phones in Japanese life.* Cambridge, MA: MIT Press.
Jenkins, H., Clinton, K., Purushotma, R., Robison, A., & Weigel, M. (2006). Confronting the challenges of participatory culture: Media education for the 21st century. An occasional paper on digital media and learning. The John D. and Catherine T. MacArthur Foundation. Available http://digitallearning. macfound.org/site/c.enJLKQN1FiG/b.208773/apps/nl/content2.asp?content_id={CD911571-0240-4714-A93B-1D0C07C7B6C1}&notoc=1. Accessed 15th May 2007.
Koutsogiannis, D. (2007). A political multi-layered approach to researching children's digital literacy practices. *Language and Education, 21*(3), 216–231.
Kress, G. (2003). *Literacy in the new media age.* London: Routledge.
Kress, G., & van Leeuwen, T. (1996). *Reading images: The grammar of visual design.* London and New York: Routledge.
Kress, G., & van Leeuwen, T. (2001). *Multimodal discourse: The modes of contemporary communication.* London: Arnold.
Leikin, R. (2005). Europe's Angry Muslims. *Foreign Affairs,* July/August 2005, Reproduced by the Council of Foreign Relations Available http://www.cfr.org/publication8218/. Accessed 1st May 2007.
Lister, M., & Wells, L. (2001). Seeing beyond belief: Cultural studies as an approach to analysing the visual. In T. van Leeuwen & C. Jewitt, C. (Eds.), *Handbook of visual analysis* (pp. 61–91). London: Sage.
Scollon, R., & Scollon, S. (2003). *Discourses in place: Language in the material world.* London: Routledge.
Sturken, M., & Cartwright, L. (2001). *Practices of looking: An introduction to visual culture.* New York: Oxford University Press.
TripleH. http://www.flickr.com/photos/38812427@N00/435910645/in/pool-83295436@N00/. Accessed 30th April 2007.
van Leeuwen, T. (2001). Semiotics and iconography. In T. van Leeuwen & C. Jewitt (Eds.), *Handbook of visual analysis* (pp. 92–118). London: Sage.
van Leeuwen, T., & Jewitt, C. (Eds.). (2001). *Handbook of visual analysis.* London: Sage.

12 Productive Pedagogies
Play, Creativity and Digital Cultures in the Classroom

Jackie Marsh

INTRODUCTION: THE TECTONIC PLATES OF HOME AND SCHOOL

This chapter takes a rather different route from previous chapters in this volume; in it, I wish to explore ways in which the introduction of aspects of children's digital cultures into the classroom can promote 'productive pedagogies' (Lingard, Ladwig, Mills, Bahr, Chant, et al., 2001) in specific kinds of contexts. By 'specific kinds of contexts', I refer to classroom spaces in which curriculum and pedagogy are respectful of children's agency, attuned to their cultural capital (Bourdieu, 1977) and embedded in meaningful practices that reflect pupils' out-of-school interests and experiences.

It is important to consider the ways in which children and young people's digital cultures can be incorporated into classrooms for a number of reasons, which include issues of relevance and significance. These arguments have been well-rehearsed elsewhere and so will not be dwelt upon here (Alvermann, Moon, & Hagood, 1999; Dyson, 2002; Marsh & Millard, 2005). In this volume, we have been introduced to rich tapestries of children and young people's out-of-school digital practices which reflect something of the complexity and intensity of literacy practices in a new media age. If classrooms do not begin to learn from and adapt some of these practices, they may become further ossified and locked into pre-21st century modes of communication (Kress, 2003). Despite the burgeoning of projects based in schools in which digital media are prevalent (Bearne et al., 2007), I would argue that there is a growing divide between home and school literacy practices. Butler and Robson (2001), in an analysis of the way in which social class operates in neighbourhood change in London, described different social class groups as tectonic in nature in the way in which the various groups they studied rarely integrated in social and cultural institutions. They suggested that: 'Social groups or "plates" overlap or run parallel to one another without much in the way of integrated experience' (Butler & Robson, 2001, p. 2157). I propose that this metaphor can be meaningfully applied to the way in which school and home contexts operate in the digital age. Whilst not ignoring the way in which children and young people transfer practices and knowledge across the various spaces they inhabit (Bulfin & North, 2007), the tectonic plates of home and school appear to be moving in very different directions in relation to digital literacy practices. This can be characterised across numerous digital literacy practices, particularly the use of Web 2.0 social networking sites and practices, such as online virtual worlds.

Virtual worlds have become increasingly popular with primary-aged children over the last two years and sites that are frequently mentioned by children and parents include *Club Penguin*, *Webkinz*, *Neopets* and *Barbie Girls*. The worlds differ in terms of their affordances, but sites such as *Club Penguin* and *Barbie Girls* enable children to create and dress-up an avatar, decorate their avatar's home, buy and look after pets and play games in order to earn money to purchase items for their avatars and homes. Both of these virtual worlds also enable interactive chat that is tightly controlled and monitored in order to allay parental concerns regarding Internet safety. This seems to be a successful strategy, as there are numerous sites across the web in which parents state that they feel comfortable with the safety measures in place, as this typical post attests:

> i let my kids useclub penguin and i think it is perfectly safe
>
> i read through all the parents bit and privacy and safety and it is completely safe
>
> it also teaches your kids the rules of chatting online and i would recommend it to every one else
>
> (Posted by: sophie at February 20, 2007, 01:22 p.m.[1])

This parent's desire for their children to learn the practices associated with social networking is one shared by many others. In a recent report, the National School Boards Association (NSBA, 2007) in the United States surveyed 1,039 parents and stated that the majority of parents held positive views regarding the educational potential of social networking sites. Similarly, in the 'Digital Beginnings' study in the United Kingdom, parents demonstrated positive attitudes towards the role of new technologies in their children's lives (Marsh, Brooks, Hughes, Ritchie, & Roberts, 2005).

Although these virtual worlds are ostensibly aimed at 8–14-year-olds, inevitably there are reports of 5- and 6-year-olds using them. These sites offer children opportunities for engaging in online social networking with others and this has been the main focus for those adults concerned about such practices, who worry that it may lead to a decline in literacy practices. However, a survey of the sites indicates that literacy is deeply embedded in the virtual worlds, as children have opportunities to write online messages, read others' messages, read catalogues, magazines, newspapers and instructions. The literacy skills, knowledge and understanding these virtual worlds can foster include:

- Reading skills and strategies including: word recognition (e.g., the vocabulary choices offered in 'safe chat' mode, in which children can chat to others using a set of already defined words and phrases; instructions; in-world environmental text), comprehension, scanning text in order to retrieve appropriate information, familiarity with how different texts are structured and organised, understanding of authors' viewpoint, purposes and overall effect of the text on the reader;
- Writing skills and strategies including: spelling, punctuation, syntax, writing using and adapting a range of forms appropriate for purpose and audience, using language for particular effect;
- Writing for known and unknown audiences; and
- Using text to negotiate, collaborate and evaluate.

In addition, children develop skills across the visual, gestural and aural modes as they juxtapose words with image, move avatars across the screen and listen to in-world oral texts. There are aspects of these sites that deserve further investigation, such as the restrictive representations of emphasised femininity in *Barbie Girls* and the promotion of commodity purchasing as a key activity in both *Barbie Girl* and *Club Penguin*. Just as forms of capital (Bourdieu, 1980/1990) operate in virtual worlds inhabited by adults, such as *Second Life*, the child-orientated worlds are also shaped by the flows of social, economic and cultural capital. In that sense, play in these worlds can be seen as 'a privileged and generative site for developing the agency of cultural producer, or "worker"' (Ito, 2007, p. 2). Nevertheless, it is clear that these sites, along with other social networking sites and practices, have much potential for developing a range of skills, knowledge and understanding in relation to digital literacy.

However, despite the burgeoning popularity of virtual worlds and other Web 2.0 sites for this age group, and their potential for developing literacy skills, primary schools in general have yet to recognise their potential. Indeed, firewalls implemented by many local authorities prevent teachers from exploring these worlds and other social networking sites in school. Even in cases in which authorities have enabled schools to be more adventurous, there is no guarantee that these sites will be used in schools in ways that replicate home uses. Merchant (2007), for example, reports on a network of primary schools in England that created a virtual world for children using Active Worlds, but then recounts how traditional practices were embedded within the design of the worlds and the use made of them by teachers. This is a phenomenon replicated across most of children's out-of-school digital literacy practices. Despite this, there is a growing focus on ways in which classrooms can incorporate Web 2.0 practices and studies are emerging that can illuminate the pedagogical possibilities offered by them. In this chapter, I will focus on the popular Web 2.0 practice of blogging in order to explore the potential for learning when popular digital forms and practices cross the home–school divide.

Blogs, or weblogs, consist of online posts which authors place on to their blog pages, with the most recent entries appearing first. Blog entries can concern a number of things, such as authors' everyday lives (similar to diary entries), news, politics and entertainment, amongst other categories. Blogs enable commentators to make comments on the entries, although this facility can be disallowed by blog authors. This commenting facility means that blogs offer rich potential for social exchanges and collaboration (Davies & Merchant, 2007; Lankshear & Knobel, 2006). Blogs can also contain images and links to other sites which enable bloggers to embed video and podcasts (audio files) into their blogs. The practice of blogging has become widespread, with estimates that the number of blogs is currently in excess of 60 million, although the number of active blogs may be much fewer than that. In this chapter, one teacher's use of blogging will be explored in order to identify how such practices can inform the development of 'productive pedagogies' (Lingard et al., 2001) in contemporary classrooms.

The notion of 'productive pedagogies' is not harnessed to a normative understanding of 'productive' that is often embedded in educational discourses, that is, that of the learner as needing to make a productive contribution, normally economical, to society. Instead, in this context, 'productive' is allied to concepts of social justice (Lingard, 2005). The productive pedagogies model was one developed in The Queensland School Reform Longitudinal Study (Lingard et al., 2001) in which observational data from approximately 1,000 primary and secondary classrooms were used to map pedagogical practices across a number of elements. These elements were derived from an extensive review of related literature and were drawn together across four dimensions: intellectual quality, connectedness, supportive classroom environment and engagement with difference. Together, these dimensions constitute productive pedagogies that can facilitate social justice in schools in that they ensure learner agency, relevance and challenge. I will focus on one of these dimensions in relation to digital literacy pedagogy—connectedness. Whilst Lingard et al. (2001) suggest that the four dimensions should not be seen in isolation, given the importance of connecting in-school and out-of-school digital literacy practices (Lankshear & Knobel, 2006) it would seem pertinent to focus in depth on this dimension here.

The following case study is based on the work of one teacher who works in a primary school in the north of England, Peter Winter. Peter is an ICT specialist who works with all of the classes in the primary school over the course of a week in a specially-equipped ICT suite which contains computers and an interactive whiteboard. The school is situated in an area of socioeconomic deprivation and serves a predominantly white, working class community. Peter was keen to identify ways in which products and services available on Web 2.0 could be used to support literacy teaching and learning and 2 years ago introduced a Year 4 class, with children aged 8 and 9, to blogging. He collaborated with a teacher in the United States on the project, and both teachers planned to enable children to develop blogs based on a mutually agreed topic, that of dinosaurs. The aim of the project was to identify the affordances of blogs for collaborative work between children located in different countries. In the subsequent year, when children were in Year 5, they developed additional blogs in response to other topics of study. In this chapter, I will focus the blogging practices of the children when they were in the Year 4 class, drawing from field notes that were made on visits to the school in which I observed Peter's pedagogical approaches. The field notes were initially inductively coded (Strauss, 1987) and then deductively coded against the elements in the dimension 'connectedness'. It is the second level of analysis that I draw from in this chapter. In addition, the chapter incorporates extracts from the blogs themselves. The aim of my analysis is to identify how pedagogical approaches that make space for digital play and creativity can be developed in ways which connect to the everyday literacy lives of pupils outside of the classroom.

PRODUCTIVE LITERACY PEDAGOGIES

Developing a challenging and relevant literacy curriculum means paying attention to the way in which the work undertaken in classrooms relates in meaningful ways to the world outside of schools. Lingard et al. (2001) suggest that the 'connectedness' dimension that they identified as a significant aspect of a productive pedagogy involves the following elements:

- Knowledge integration. (Does the lesson integrate a range of subject areas?)
- Background knowledge. (Are links with students' background knowledge made explicit?)
- Connectedness to the world. (Is the lesson, activity, or task connected to competencies or concerns beyond the classroom?)
- Problem-based curriculum. (Is there a focus on identifying and solving intellectual and/or real-world problems?)

In the following section, I will examine these factors in turn.

Background Knowledge

One of the main elements of the connectedness dimension is the focus on how far work undertaken in the classroom is connected to competences and concerns beyond the classroom. The majority of children in the Year 4 class that Peter worked with initially were competent and enthusiastic users of a range of new technologies outside of school. Peter was aware of this and keen to ensure that the curriculum drew from the students' expertise. Whilst none of the students had previously developed a blog at that

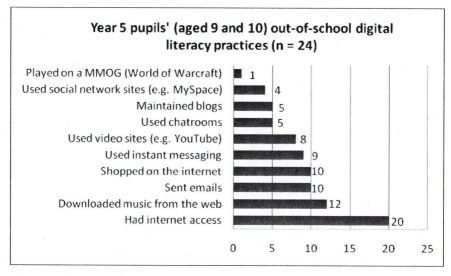

Figure 12.1 Internet use.

point, many were familiar with the use of the Internet for collaboration and exchange of ideas through software such as Instant Messenger. However, when 24 of the children in this class completed a survey of their out-of-school use of technologies a year later, in Year 5, five children indicated that they now kept their own blogs out of school (see Figure 12.1).

This is interesting in relation to the question of whether or not the appropriation of out-of-school practices for schooled purposes is off-putting for pupils; in this case, the blogging project did not appear to deter five of these children from creating blogs at home.

Numerous studies have indicated how pervasive digital media are in young children's lives (Burnett & Wilkinson, 2005; Livingstone & Bober, 2005; Ofcom, 2006; Valentine, Marsh, & Pattie, 2005). Even very young children now have extensive experience with technologies such as computers, console games, DVD players and mobile phones (Marsh, 2005; Marsh et al., 2005). Figure 12.1 indicates that this group of children were well immersed in many of the social networking practices prevalent in contemporary society such as blogging and the use of social network sites such as MySpace and Bebo. Therefore, when the pupils were introduced initially to the concept of a blog, they had no difficulties understanding how it worked and very soon acquired the skills and knowledge needed to post to it and send comments.

Peter informed the children that they could use the blog to disseminate any aspects of the dinosaur topic that they felt to be important. As a consequence of the learner autonomy embedded in this pedagogical model, the blog soon wove together a range of genres—informative texts, reports,

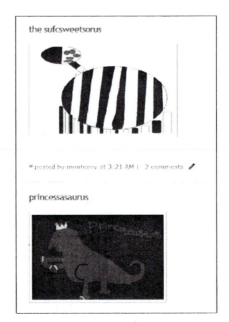

Figure 12.2   Blog post 1.

Figure 12.3   Blog post 2.

narratives, lists; replicating the textual mix pupils meet in online worlds outside of schools. Inevitably, given the high level of ownership of the task enjoyed by the pupils, references to the pupils' popular cultural interests permeated the posts. Figure 12.2, for example, contains a picture of a dinosaur in the colours and stripes of a local football team, to which classmates posted enthusiastic, supportive comments about the ability of the team. Figure 12.3 includes a post in which a child uploaded a photograph of a favourite dinosaur character in the film *Toy Story*.

The affordances of blogs mean that they are ideal formats for displaying aspects of one's identity. As Carrington (2006) notes:

> . . . these texts are signposts of the kinds of practices with technology and text that may be socially useful in developing and displaying self-narratives—layered, networked texts, multimodality, the continuous and conscious slide between online and offline. (p. 11)

The blogs enabled children to reference popular texts and artefacts that represented aspects of their identities and thus blurred their online and offline worlds (Davies, 2005; Merchant, 2004). Similarly, the blogs were representative of a range of multimodal texts children meet outside of the classroom and so connected intimately to their everyday literacy practices. As Carrington (2006) notes of blogs, print may act in a primary or subsidiary role and, 'Significant or even core content may well be lodged in a visual image or in an audio file or as a collage that draws upon all available modes rather than in the print component itself' (Carrington, 2006, p. 6). This replicates children's interactions with a wide range of digital texts such as moving image texts or computer games. In this way, the blogging project drew extensively on children's background knowledge and out-of-school practices.

Connectedness to the World

A further aspect of the 'connectedness' dimension relates to the way in which lessons are 'connected to competencies or concerns beyond the classroom' (Lingard et al., 2001). In this case, Peter used the work on blogging to reinforce aspects of Internet safety, which he was keen to develop in order that children could develop good practice when using the Internet outside of school.

There is currently much interest in this area, with the Byron Review (Department for Children Schools and Families, 2007) in the United Kingdom focusing on developing guidance for parents and others on keeping safe online. This review has been commissioned in response to growing public concern regarding children and young people's Internet use. As technological developments intensify the pace of change in society at large, there is a corresponding proliferation of moral panics in relation to children's use of these technologies. For example, in the United Kingdom last year a letter was sent to a national broadsheet, signed by over 100 early years specialists, academics and practitioners, which outlined a series of concerns about contemporary childhoods. The letter included the following paragraph:

> Since children's brains are still developing, they cannot adjust—as full-grown adults can—to the effects of ever more rapid technological and cultural change. They still need what developing human beings have always needed, including real food (as opposed to processed 'junk'), real play (as opposed to sedentary, screen-based entertainment), first-hand experience of the world they live in and regular interaction with the real-life significant adults in their lives. (Abbs et al., 2006)

This is misleading on a number of accounts. There is a false juxtaposition here that sets up engagement with technologies and 'real' play as oppositional, as explored in Cross's chapter in this volume. In addition, it should be noted that screen-based entertainment is not exclusively sedentary (Marsh et al., 2005). Further, in March 2006, David Willets, the Conservative Shadow Education Secretary, set up a formal inquiry into 'Lost Childhoods' in England, following a UNICEF (2006) report that indicated that the United Kingdom ranked bottom in a well-being assessment of children in 21 industrialised countries. Rather than questioning the methodology utilised in the UNICEF study, this knee-jerk reaction typified a range of responses to the current climate, which included the emergence of a book with the rather provocative title, '*Toxic Childhood*' (Palmer, 2006).

These reactions were symptomatic of the frequent, negative reactions some adults express towards changing childhoods. There is no doubt that contemporary childhoods are being transformed, with social and cultural changes taking place that have significant implications for the teaching and learning of literacy. There is also no doubt that the nature of public spaces is changing for children and young people. Many children and young people are involved in social networking sites such as *Bebo* and *MySpace* (Dowdall, this volume) and this is potentially confusing and alienating for teachers who grew up with very different experiences of engagement with known and unknown audiences. Teachers are anxious about safety aspects of the Internet (Demos, 2007) and yet in a recent U.S. study conducted by the National School Boards Association (NSBA, 2007), only 0.08% young people in the survey reported meeting people they had met over the Internet without their parents' permission. This is not to minimise the concerns expressed by teachers, but suggests that instead of becoming overprotective in online spaces, we need to engage with young learners as they develop further their critical capacities and begin to make judgements about, for example, which aspects of their identities they share with which audience(s) at any one time.

The work of children in the blogging project featured in this chapter was informed by the school's Internet policy, which encouraged the use of critical literacy practices in the choice and use of online sources. In this way, the children were developing important skills that they would use outside of the classroom, skills relevant to an information-economy (Lankshear & Knobel, 2006) and skills that could enable them to foster safer use of the Internet. In this work, Peter frequently drew children's attention to the way in which the project linked to home Internet practices, reminding them of the strategies they already knew and were using both in school and out-of-school:

> Peter began the session by reminding children about the 'rules' they were already using at home and school when they went on the web i.e. they should not use their real name to identify themselves, they should never give out their real name and address to anyone over the web, if they came across untoward material, they should click off it immediately. Peter had placed the school's Internet policy near each of the computers and drew children's attention to it, but stressed that they already knew the policy. The children all greeted this remark with equanimity—there were no comments or questions, they simply moved to the computers with little fuss. None of the children referred to the policy throughout the session, but all adhered to it. This was obviously well-embedded practice for them. (Field notes, May 18, 2006)

A further area of consideration here is the issue of copyright. As Web 2.0 further dissolves the boundaries between production and consumption and celebrates a 'remix' culture (Lankshear & Knobel, 2006) in which 'produsage' (Bruns, 2006) abounds, anxieties around copyright and the line between collaboration and collusion proliferate. Peter had a number of concerns about this as the children began to mine the web for material to place on their blogs, but instead of becoming paralysed by fears surrounding this issue, he encouraged the children to consider the nature of their sources and acknowledge them where appropriate, or link directly to their web source. There are no simple solutions to an area that confounds many copyright lawyers and as this field develops, teachers need to be part of the dialogue about the nature of intellectual property in the digital age.

Knowledge Integration

The work on this project was also strong in relation to another element of the 'connectedness' dimension, the integration of knowledge across subject areas. The dinosaur project enabled children to draw from and extend knowledge across a range of areas, including science, history, geography, literacy, technology and art. The affordances of the technology meant that this cross-curricular approach was an integral part of the work, given the

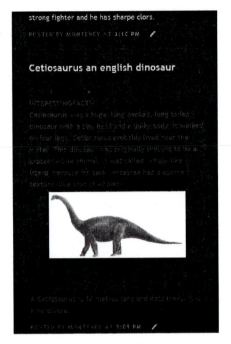

Figure 12.4   Blog post 3.

range of online sources the children could access and the variety of software they used throughout the project. The integration of knowledge occurred across blog posts, as popular cultural references were placed alongside posts in which children exchanged scientific and/ or historical information about dinosaurs, in this case the Cetiosaurus (see Figure 12.4).

The blogging project also enabled children to integrate knowledge about literacy with technology in, for example, the development of keyboard skills, but also fostered digital literacy skills in which literacy and technology were integrated. Throughout the blogging project, the boundaries between subjects could be seen to be weakly classified (Bernstein, 2000), thus enabling the integration of horizontal and vertical discourses (Bernstein, 2000) and the creation of 'third space', or recontextualised, knowledge (Moje et al., 2004).

Problem-based Curriculum

The fourth element of the 'connectedness' dimension, the focus on identifying and addressing intellectual/ real-world problems, was also a central element of the blogging project. For example, Peter encouraged the children to explore the blogging software in order to identify its affordances and constraints:

Two children had worked on their drawing of the dinosaur. They were unsure of how to post it on the blog, and so Peter encouraged the pupils to save it to their folder on the desktop, then log on to 'Blogger' and try and work out which icon they used to upload the picture. The boys took some to time to work out how to save their picture and then log on to 'Blogger', but eventually they did so, identified the correct icon, uploaded their picture and posted a title. (Field notes, June 6, 2006)

There were other examples of children being challenged to address a range of technical and intellectual problems as they created short animated films for the blog using software they were previously unfamiliar with, investigated the process of linking sites to the blog or worked out how to find out certain information they required about dinosaurs drawing from a range of online sources.

The work discussed so far indicates the way in which blogging can promote a pedagogical approach which fosters creativity, although of course it is not necessarily the case that all classroom blogging projects do so. What is significant in this example is the way that the blogging was predicated on the personal choices of children in terms of what they choose to write posts about and enabled them to draw extensively on out-of-school knowledge and practice. In this way, issues of voice, identity and agency rise to the fore. As Lewis suggests:

> New literacies tend to allow writers (users; players) a good deal of leeway to be creative, perform identities, and choose affiliations within a set of parameters that can change through negotiation, play and collaboration. (Lewis, 2007, p. 231)

In the next section, I review the way in which both play and creativity are embedded within the blogging practice and reflect on the pedagogical implications.

PLAY AND CREATIVITY IN CLASSROOM BLOGGING

As many of the chapters in this volume indicate, both 'play' and 'creativity' are nebulous concepts which defy precise definition. In this instance, I shall use them as adverbs in order to discuss the processes in which children engaged as they blogged, processes which featured creative and playful aspects. This need not be construed as a wish to sidestep the whole debate, as it draws on previous discussions of play, such as that by Millar:

> Perhaps play is best used as an adverb; not as a name of a class of activities, nor as distinguished by the accompanying mood, but to describe how and under what conditions an action is performed. (Millar, 1968, p. 21)

Throughout the blogging project, children were playfully engaged with each other. For example, they used the commenting feature of the blogging software to interact with friends. During the dinosaur project, the children decided that they needed a separate blog linked to the original 'Dinoblog' which would enable them to share information about themselves with the American class and so an 'All about us' blog was created. John posted an entry on it (Figure 12.5) and then Sam posted a comment on John's post (Figure 12.6) even though he was sat in the same locality and could have vocalised his response.

Indeed, comments were often used to perform friendships, as children posted comments to each other on blog posts and then went over to tell each other that they had just done this. This playful engagement with each other through the blogs was similar to the way in which many adult bloggers operate, with comments often being posted by people known to the bloggers (Davies, this volume; Davies & Merchant, 2007). Throughout the project, there were numerous examples of how children played out friendships across the blogs and, frequently, pictures were chosen in order to impress peers and invite interaction. At times, children named each other in their entries in order to invite exchange (Figures 12.7 and 12.8):

Figure 12.5   Blog post 4.

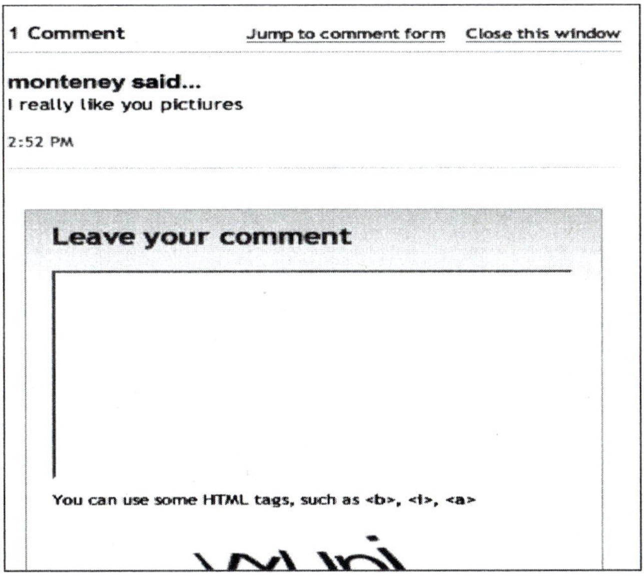

Figure 12.6   Blog post 5.

This use of classroom activities to perform friendships and play out discourses threading into classroom life from playground culture is not peculiar to online practices such as blogging, of course, as Dyson's illuminating case studies of children's paper-based writing practices demonstrate (Dyson, 2002). However, the commenting facility in 'Blogger' did mean that written tokens of friendship were on display for all to see and were thus permanent markers of shifting patterns of relationships. The comments invited performance of identities in public spaces and led to playful encounters with texts and images as children constructed both themselves and others in the practice of blogging.

In addition to these instances of play, the blogging project fostered a range of creative activities. Children drew on a variety of web sources, as well as their own productions, to mix and match in the production of new texts, reflecting the type of creative activities already well embedded in a 'DIY Internet' (Sharp, 2006) culture. Figure 12.9, for example, is a still from a film in which pupils mixed photographs they had staged using plastic dinosaurs with text and dramatic music, in an attempt to create a film in the style of *Jurassic Park*. The film was then posted on YouTube and linked to the blog.

In addition to films created by the sequencing of still images, the children also produced animations and a film which contained spliced extracts from one of their animated films with a short film of a volcano erupting in order to represent, as the film's title suggested, the 'End of the dinosaurs'.[2]

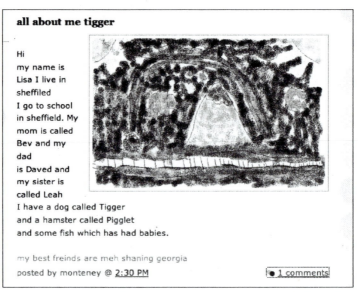

all about me tigger

Hi
my name is
Lisa I live in
sheffiled
I go to school
in sheffield. My
mom is called
Bev and my
dad
is Daved and
my sister is
called Leah
I have a dog called Tigger
and a hamster called Pigglet
and some fish which has had babies.

my best freinds are meh shaning georgia
posted by monteney @ 2:30 PM   ● 1 comments

Figure 12.7   Blog post 6.

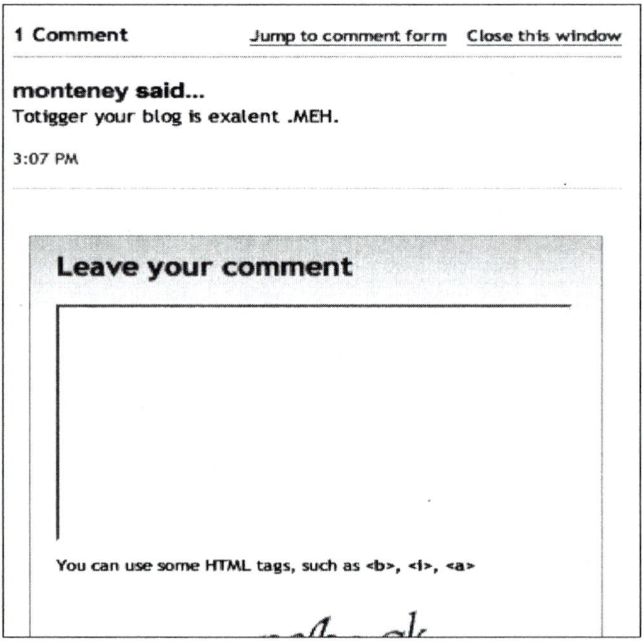

1 Comment   Jump to comment form   Close this window

monteney said...
Totigger your blog is exalent .MEH.

3:07 PM

Leave your comment

You can use some HTML tags, such as <b>, <i>, <a>

Figure 12.8   Blog post 7.

The giant dinosaur moves through the lush undergrowth. "I must find food soon or I will starve to death."

Figure 12.9   Still picture from 'DinoWars 2'.

This type of 'mash-up' (Shiga, 2007) activity is becoming more prevalent in out-of-school digital media practices, as Willett's chapter in this volume outlines. The value of embedding this type of media creativity in the schooled literacy curriculum is that pupils can be encouraged to enhance their skills further, developing the quality of critical reflection on their work which may then extend to their out-of-school productions. Simply because children engage in out-of-school digital media production does not mean that they necessarily develop the range of skills and practices required for further development and refinement. In addition, given that some children do not have extensive access at home to new technologies that enable online production, it is important that schools offer all pupils opportunities to develop the kind of networking and 'DIY Internet' (Sharp, 2006) skills that are going to become so important for the future.

The level of playfulness and creativity permeating the blogging project need not be confined to the use of digital media in the curriculum and that is not the argument I am attempting to make here. Rather, the pedagogical strategies deployed by Peter were the key to ensuring that this project was so successful in fostering independent learning. In particular, the emphasis on one of the dimensions identified in the 'Productive Pedagogies' project

(Lingard et al., 2001), that of 'connectedness', was central to motivating pupils and drawing on their digital 'cultural capital' (Bourdieu, 1977).

CONCLUSION

This example of the use of contemporary digital literacy practices in one primary classroom has been used to identify ways in which such work can contribute to the development of productive pedagogies that offer all children an opportunity to build on out-of-school knowledge and practices. The playfulness and creativity which is inherent in much of children's engagement with digital media outside of the classroom can be incorporated into the curriculum if the pedagogical approach adopted is one that fosters pupils' agency and facilitates their critical engagement with a wide range of multimodal texts. There are a number of features of this kind of work which are crucial to its success, all of which can be identified in this case study. The first is, as suggested previously, that classroom activities should recognise children's digital 'habitus' (Bourdieu, 1980/1990) and enable them to draw on the knowledge and practices they develop in their engagement with new technologies in the home. Second, pedagogical practices should foster children's agency and enable them to make real choices about form, purpose and audience in text production. Third, these activities should enable identity work, promoting ways of children forging, reflecting on and performing identities in classroom contexts. Fourth, we need to engage with the opportunities offered by Web 2.0 social networking sites for promoting collaboration and the development of productive learning communities that can span in- and out-of-school spaces. Finally, classroom activities that enable children to become 'text bricoleurs' (Lankshear & Knobel, 2004) in which they weave together familiar and unfamiliar texts in ways which promote new knowledge and understanding are central to a productive pedagogy model. These kinds of pedagogical practices are not simply necessary for promoting particular types of learning; they are crucially important to ensuring that schooling remains relevant and meaningful in an age in which play and creativity are writ large in the digital cultures that surround us.

NOTES

1. Posted on *Business Week* blog at: http://www.businessweek.com/careers/workingparents/blog/archives/2006/09/while_moms_away.html
2. See the blog at http://dinoproject.blogspot.com/

REFERENCES

Abbs, P. et al. (2006). Modern life leads to more depression among children. Letter to the Daily Telegraph, 12th September, 2006. Retrieved August 10,

2007, from from http: <http://www.telegraph.co.uk/news/main.jhtml?xml=/news/2006/09/12/njunk112.xml>

Alvermann, D., Moon, J. S., & Hagood, M. C. (1999). *Popular culture in the classroom: Teaching and researching critical media literacy.* Newark, DE: IRA/NRC.

Bearne, E., Clark, C., Johnson, A., Manford, P., Mottram, M., & Wolstencroft, H. et al., (2007). *Reading on screen.* Leicester, England: UKLA.

Bernstein, B. (2000). *Pedagogy, symbolic control and identity: Theory, research, critique* (Rev. ed.). Oxford, UK: Rowan & Littlefield.

Bourdieu, P. (1977). *Outline of a theory of practice,* Cambridge, UK: Cambridge University Press.

Bourdieu, P. (1990). *The logic of practice* (R. Nice, Trans.). Cambridge: Polity Press. (Original work published in 1980).

Bruns, A. (2006). Towards produsage: Futures for user-led content production. In F. Sudweeks, H. Hrachovec, & C. Ess (Eds.), *Proceedings: Cultural Attitudes Towards Communication and Technology 2006* (pp. 275–284). Perth: Murdoch University, Australia. Retrieved August 22, 2007, from http: <http://snurb.info/files/12132812018_towards_produsage_0.pdf>

Bulfin, S., & North, S. (2007). Negotiating digital literacy practices across school and home: Case studies of young people in Australia. *Language and Education, 23*(3), 247–263.

Burnett, C., & Wilkinson, J. (2005). Holy Lemons! Learning from children's use of the Internet in out-of-school contexts. *Literacy, 39,* 158–165.

Butler, T., & Robson, G. (2001). Social capital, gentrification and neighbourhood change in London: A comparison of three south London neighbourhoods. *Urban Studies, 38*(12) 2145–2162.

Carrington, V. (2006, May). *Texts, fugue, digital technologies.* Paper presented at Daiwa Foundation sponsored UKLA/University of Nara Seminar.

Davies, J. (2005). Nomads and tribes: Online meaning making and the development of new literacies. In J. Marsh & E. Millard (Eds.), *Popular literacies, childhood and schooling* (pp. 160–176). London: Routledge.

Davies, J., & Merchant, G. (2007). Looking from the inside out: Academic blogging as new literacy. In M. Knobel & C. Lankshear (Eds.), *A new literacies sampler* (pp. 167–197). New York: Peter Lang.

Demos. (2007). *Their space—education for a digital generation.* Retrieved August 20, 2007, from http: <http://www.demos.co.uk/files/Their%20space%20-%20web.pdf>

Department for Children, Schools and Families. (DCSF, 2007). *The Byron Review: Children and New Technology.* Retrieved November 10, 2007, from http: <http://www.dfes.gov.uk/byronreview/>

Dyson, A. H. (2002). *Brothers and sisters learn to write: Popular literacies in childhood and school cultures.* New York: Teachers College Press.

Ito, M. (2007). *Engineering play: Children's software and the cultural politics of edutainment.* Retrieved August 28, 2007, from http: <http://www.itofisher.com/mito/EngPlay.pdf>

Kress, G. (2003). *Literacy in a new media age.* London: Routledge.

Lankshear, C., & Knobel, M. (2004). Text-related roles of the digitally 'at home.' Paper presented at the American Education Research Association Annual Meeting, San Diego, April 15, 2004.

Lankshear, C., & Knobel, M. (2006). *New literacies>: Everyday practices and classroom learning* (2nd ed.). Maidenhead, Berkshire: Open University Press.

Lewis, C. (2007). New literacies. In M. Knobel & C. Lankshear (Eds.), *A new literacies sampler* (pp. 229–237). New York: Peter Lang.

Lingard, B. (2005). Socially just pedagogies in changing times. *International Studies in the Sociology of Education, 15*(2), 165–184.

Lingard, B., Ladwig, J., Mills, M., Bahr, M., Chant, D., Warry, M., et al. (2001). *The Queensland School Reform Longitudinal Study* (Vols. 1 and 2). Brisbane: Education Queensland.

Livingstone, S., & Bober, M. (2005). *UK children go online: Final report of key project findings.* London: London School of Economics and Political Science.

Marsh, J. (Ed.). (2005). *Popular culture, new media and digital technology in early childhood.* London: RoutledgeFalmer.

Marsh, J., Brooks, G., Hughes, J., Ritchie, L., & Roberts, S. (2005). *Digital beginnings: Young children's use of popular culture, media and new technologies.* Sheffield: University of Sheffield. Retrieved June 3, 2006, from http: <http://www.digitalbeginings.shef.ac.uk/ >

Marsh, J., & Millard, E. (Eds.). (2005). *Popular literacies, childhood and schooling.* London: RoutledgeFalmer.

Merchant, G. (2004). Imagine all that stuff really happening: Narrative and identity in children's on-screen writing. *E-learning, 3*(1), 341–357.

Merchant, G. (2007, July). Daleks and other avatars. Paper presented at the UKLA Conference, University of Swansea.

Millar, S. (1968). *The psychology of play.* Harmondsworth, England: Penguin.

Moje, E., Ciechanowski, K., Kramer, K., Ellis, L., Carrillo, R., & Collazo, T. (2004). Working toward third space in content area literacy: An examination of everyday funds of knowledge and discourse. *Reading Research Quarterly, 39*(1), 38–70.

National School Board Association. (NSBA, 2007). *Creating and connecting: Research and guidelines on social—and educational—networking.* Retrieved August 20, 2007, from http: <from: http://www.nsba.org/site/docs/41400/41340. pdf >

Ofcom. (2006). Media literacy audit: Report on media literacy amongst children. London: Ofcom. Retrieved August 20, 2007, from http: <http://www.Ofcom. org.uk/advice/media_literacy/medlitpub/medlitpubrss/children/children.pdf>

Palmer, S. (2006). *Toxic childhood.* London: Orion Press.

Sharp, D. (2006). Participatory culture production and the DIY Internet: From theory to practice and back again. *Media International Australia Incorporating Culture and Policy, 118*, 16–24.

Shiga, J. (2007). Copy-and-persist: The logic of mash-up culture. *Critical Studies in Communication, 24*(2), 93–114.

Strauss, A. L. (1987). *Qualitative analysis* for *social scientists.* New York: Cambridge University Press.

UNICEF. (2006). Child poverty in perspective: An overview of child well-being in rich countries. Florence, Italy: Unicef. Retrieved Augusto 10, 2007, from http: <http://www.unicef.org.uk/press/news_detail_full_story.asp?news_id=890>

Valentine, G., Marsh, J. and Pattie, C. (2005). *Children and young people's home use of ICT for educational purposes: The impact on attainment at key stages 1–4.* London: HMSO.

13 Conclusion

Muriel Robinson and Rebekah Willett

Debates concerning digital cultures suggest that new technologies are providing both benefits and risks to users. For example, as can be seen in this volume, the Internet is offering instant access to information and creating new forms of communication and community. There are clearly powerful opportunities here for self-expression, creativity and learning. However, a celebratory approach to the affordances of the Internet needs to be balanced with an examination of the power dynamics inherent in people's use of the Internet. Not least is the issue of access, where although many people might have some kind of access to the Internet, the quality of the access (e.g., bandwidth, constant access versus access through schools or libraries), the social resources required to make the most of access and the economic resources needed to maintain access vary enormously. Furthermore, the benefits of Internet access also bring new risks concerning complex issues such as trust, credibility, privacy and commercialism, as recent reports about the growing incidence of criminal activity and pornography in the virtual world, Second Life, exemplify (Wade, 2007). The aim of this book has been to suggest ways that notions of play and creativity might inform our analyses of digital cultures, to provide examples of such analyses and to engage with continuing debates about people's participation within digital cultures. The processes involved in participating in digital cultures are complex, and one of the challenges facing researchers in this area is to develop frameworks for analyses, as the authors in this volume have done, which reflect some of these complicated interactions.

Returning to the circuit of culture which we introduced in Chapter 1, analysing digital cultures through the different processes (consumption, production, representation, regulation and identity), and particularly looking at the moments when these processes connect, helps to highlight benefits and the risks as discussed above. Analysing representation, for example, might take account of both the risks and benefits of Internet cultures (e.g., times when a topic is represented democratically and inclusively or times when the reverse is true and credibility issues arise). Production of information on the Internet is connected with how information is represented (the motives of the producer, for example), and again issues of trust and credibility must be examined. Whilst seeing how all the processes on the circuit of culture might work in relation to digital cultures, the research in many chapters presented in this volume concentrates on consumption and production practices, understood in a sense which takes as central the active and creative aspects of consumption and a two-way interactive relationship between many acts of consumption and production. Here the notions of play and creativity help explain the actions of the consumer, and also help position consumption of digital technologies as productive acts. Consuming video games and then playfully acting them out is an act of consumption and production; consumers on Bebo, Flickr and YouTube are also producing specific identities for themselves through creative engagements on those sites.

To end this volume we return to the questions posed in Chapter 1 and link together the research presented in the chapters by focusing on three key questions:

- What notions of creativity are useful in our fields?
- How does an understanding of play inform analysis of children's engagement with digital cultures?
- How might school practice take account of out-of-school learning in relation to digital cultures?

These are questions which underpin much of the research outlined in this volume and they are questions that are central to future developments in this field. Educators and researchers involved in the analysis of children and young people's engagement with digital cultures need to reflect on these issues in order to ensure that the relationship between informal and formal learning spaces becomes better defined and creative and playful approaches to the use of digital technologies are developed across domains.

WHAT NOTIONS OF CREATIVITY ARE USEFUL IN OUR FIELDS?

One of the underlying premises of much of the research concerning children's digital cultures is that children are engaging with media in ways which we might call 'creative'. Focusing on 'digital cultures', rather than separating out technology or media from those cultures, implies that we are interested in the meanings that are being made from and through media. Rather than seeing meaning as inherent to a particular digital technology or media text, or children as being positioned in particular ways by technologies, the focus on culture implies that there is creative meaning negotiation occurring as children and young people engage with technologies. Here we see identity as well as meanings of texts being negotiated (text being defined in a broad sense to include images; peer to peer communication; catch phrases; and characters and plot lines from movies, novels, TV shows, games, etc.). It is not only meaning that involves creative negotiation; the texts themselves can be seen as creative expressions, as we see with designs on web pages, online videos and photographs, as well as written texts (including message boards, blogs and online fiction). Importantly, here creativity is being defined as part of everyday engagement within cultures, and this definition is specific to particular traditions (e.g., media and cultural studies).

The problematic nature of creativity as a concept and the multiple understandings which can lie behind it are significant factors in the work presented in this collection. Banaji's work helps to highlight specific uses of the term 'creativity' by analysing the different ways 'creativity' as a term and concept have been and are being used in a variety of contexts. In relation to digital technologies, we see different, often contradictory, applications of the term 'creativity'. Whilst some people (educators, researchers, authors, parents, etc.) see new technologies as encouraging and enabling young people's creativity, allowing creative expression to flourish, others blame technologies for stunting creativity. By examining philosophical and historical traditions, it is possible to see how 'creativity' is used in oppositional and political ways. For example, some traditions define creativity as unique, as something special that only certain people possess; other traditions discuss creativity as something that can be fostered and developed; whilst still others use a definition of creativity which is far more inclusive and democratic, referring to a quality that exists in all people and can be seen in everyday interactions. Returning to another theme we identified in Chapter 1, we would want to stress that in education, we sometimes see creativity connected with programmes for children who are at risk of exclusion, seeing creativity as empowering, as a way of connecting with particular cultures and combating societal problems. In schools, creative teaching and learning are referred to as connected with motivation and quality, and often creativity is discussed as something that can be measured and assessed. Furthermore, a connection is sometimes made between play, creativity and learning, and here there is fear that the increasing demands of a performance-based curriculum is making less time for play and therefore less chance for creative kinds of learning. In relation to digital technologies and learning, many researchers take a cautious approach, analysing the various factors which support or infringe on children's and young people's use of technology for what might be seen as creative purposes. The importance of examining these different rhetorics of creativity is to recognise that the term creativity, although used in everyday discourse in imprecise ways, is not a value neutral term—there are political implications for using the term in particular ways, views of cognition, culture and society which are implied in uses of the term creativity, and implications for how we as educators and researchers discuss children and young people's play with digital cultures.

As implied above, the approach throughout this volume draws on rhetorics which position creativity as democratic and ubiquitous, whilst at the same time trying to tease out the connections between play and creativity. Many of the chapters analyse ways digital cultures are being engaged with through creative productions and in which children and young people are defining and performing their identities through creative engagement with digital technologies. Dowdall's and Willett's chapters specifically analyse identity work which young people are engaging with through digital productions, negotiating meaning around the texts and images they post online and shifting and managing their on and offline identities. Davies offers us detailed and rich descriptions of the creative possibilities of online communities, from her study of preteen girls playing Babyz to adults' playful encounters on Flickr and through their blogs. She sees play and creativity as being in many ways interconnected and emphasises how playful creativity can lead to 'collaborative, participatory texts which support individuals to communicate in innovative ways' (Davies, this volume).

Carrington shows how the creative act of photography enables a new model of citizenship and a new way of sharing interpretations of key events. The iconic image with which she engages in the first part of her chapter was not carefully posed or edited but was an immediate capture of a moment in time—a model of creativity which emphasises its spontaneous and tentative nature. The taking and sharing of such an image is a creative act open to anyone with a camera phone and the capacity to upload it to a web-based network. Such a model of creativity offers great possibilities for exploring how people engage with new technologies and digital cultures in a way which extends our understanding of the tension between the local and global.

Mackey's study of readers engaging in thick play also engages with creativity, emphasising the ways in which those readers whose thick play takes them into big worlds can seek to develop the detail of such worlds and make connections between different manifestations of these worlds. This is creativity as intertextual and to a large extent collaborative, the original texts extended and reshaped in the context of the play. Similarly, Cross's study of boys' engagements with computer based games analyses the kind of thick play that Mackey refers to, creatively transforming plots, characters and spaces as they move between digital worlds and their communal offline worlds.

These examples of everyday interactions with technologies which are situated within particular cultures are usefully explained with reference to creativity. As the chapters in this volume have demonstrated, people are not just replicating the kinds of interactions that are inherent to the technologies (playing a computer game, reading a text, posting a photo or video). The interactions are complex, and one might say 'creative', as the players negotiate meaning and identities, select and transform texts to suit their purposes and engage in new understandings of technologies. As many of the contributors here acknowledge, and as seen in Banaji's discussion in particular, creativity and play are closely related concepts. However, they are not coterminous and we also here want to bring out the key issues as revealed by our contributors with regard to the particular links between play and digital cultures.

HOW DOES AN UNDERSTANDING OF PLAY INFORM ANALYSIS OF CHILDREN'S ENGAGEMENT WITH DIGITAL CULTURES?

Much of the daily discourse around play in recent years has started from a concern that digital play may be displacing 'real' play with deleterious effects. This model of displacement is familiar to us from rhetorics around the dangers of television and other 'new' media (see, e.g., Winn, 1985) around the potential for the new activity to limit time for something seen as more worthy, such as private reading time. We have both argued elsewhere in detail about the fallacy of these arguments, which can be traced back to earlier concerns that the novel was a dangerous and foolish occupation displacing more serious reading, or that theatre was similarly damaging to our cultural capital (Robinson, 1995; Willett, 2005). As Marsh shows in her chapter, the recent worries about 'real' play being displaced offer what she sees as a false juxtaposition, setting up engagement with technologies and 'real' play as oppositional. Several of the contributors to this volume, by offering detailed and thoughtfully analysed accounts of digital play which are grounded in a deeper understanding of the complexity of play, go some way to providing an alternative and more theorised account of the ways in which playful digital encounters can be seen as a part of a continuum of play experiences.

Play, like creativity, is a very difficult concept to pin down. As we have discussed in our introduction to this volume, rhetorics surrounding play are used to position players and their actions in particular ways. Several authors in this volume (Cross, Pelletier, Willett) apply Sutton-Smith's (1997) analysis of the rhetorics of play to discuss the ways children's play with digital media is positioned in public discourse, research and educational settings. This is not to say that play is not a useful concept through which to analyse digital cultures. Caroline Pelletier shows us how children themselves define play through complex processes of negotiation. Here, the meaning of play is constructed through social interactions with reference to individual identities. Similarly, Willett's analysis of teenage boys' play with media texts shows how meaning is negotiated and connected with the performance of identity. Using Winnicott's (1971) notion of 'third space', Willett analyses the ways in which play offers spaces to explore the meaning of cultural objects, contest dominant discourse and position players in powerful ways.

Marsh suggests that we can also develop our understanding by taking Millar's (1968) approach of using play as an adverb—looking, in other words, at examples of people behaving playfully. Marsh's own account of children's playful engagement with their blogging activity resonates well with the ways in which the children and adults discussed by Davies and others demonstrate playfulness in their interactions.

Davies is one of the contributors here who shows how such playful behaviour transcends the digital/real encounter. The people sharing digital images through Flickr meet to create further images, and her academic colleagues develop playful language use online which enriches and influences their behaviour offline. Nor is such cross-media play confined to adults. Cross's very powerful account of how a group of children use their digital play experiences in their very physical outdoor play makes it clear that these two sets of play experiences are by no means in competition with each other but coexist and enrich each other. From this deeper understanding of the connections between digital and real-world play, Cross draws important conclusions both about the importance of access to appropriate physical spaces for real world play and about the digital divides that can be created if children do not have access both to such physical spaces and to a range of possibilities for digital play.

Carrington's work on citizenship makes an equally powerful case here for play as a means to engage people in big questions around the global/local debates. She argues that digital literacy skills

> will allow them to be creative, compliant, playful and transgressive, and to take their place in the participatory culture that Jenkins describes—a culture that will operate on and off line and all the spaces and places in between. (Carrington, this volume)

Transitions across places and spaces become particularly acute when exploring the kinds of digital practices and related learning that occur in school and out-of-school settings. It is to this last question that we now turn in our reflections on how authors in this book have addressed the complex relationship between informal and formal learning in homes, communities and schools.

HOW MIGHT SCHOOL PRACTICE TAKE ACCOUNT OF OUT-OF-SCHOOL LEARNING IN RELATION TO DIGITAL CULTURES?

Many of the chapters in this book analyse engagements with digital cultures which are occurring outside of schools. Although the focus of this book is not on learning or pedagogy, it is clear from the analyses presented here that young people are engaged in learning which involves content and styles that sometimes complement and at other times contrast with those found in schools, and so, as researchers with a particular concern with the ways in which children and young people learn, it feels right and proper to address the implications for school learning which emerge from this volume. Research is showing that whilst online cultures build on traditional skills (literacy, research, critical analysis), specific new media literacies are developing through online participation (Jenkins, Clinton, Purushotma, Robison, & Weigel, 2007). In their white paper on media literacy, Jenkins et al. argue that, rather than centring on technological skills, new media literacy involves 'a set of cultural competencies and social skills' (p. 4). Jenkins et al. identify 11 new skills associated with online social environments, including appropriation, multitasking, collective intelligence, judgment, networking and negotiation. Several chapters in this volume provide evidence and analyses of these 'cultural competencies and social skills'. Davies maps informal and playful learning within a range of online communities, showing how the collaborative aspects of this shared meaning-making offer individuals the necessary competencies to participate in many different cultures, both online and off. Dowdall also looks at online and offline interactions, focusing on one girl's negotiation and management of social identity. The competencies Dowdall analyses concern self-representation through digital texts. Here we see complex ways that young people are positioning themselves in relation to peers. Importantly, Dowdall highlights how her case study's online communication is complex and sometimes contradictory, drawing on a wide variety of resources but also located in a local peer group.

The resources young people draw on in their digital play are the focus of several chapters, and in assembling this volume we have included analyses of the commercial aspects of these resources. With digital play increasingly being seen as sites for marketing, researchers are asking questions about young people's engagements with and understanding of commodified environments. These are complex spaces, particularly in light of new technologies, new forms of marketing and increasingly globalised networks. In this volume, de la Ville and Durup suggest ways that media industries are contributing to children's cultural worlds, creating meaning around global texts through various intertextual references, and Willett analyses young people's transformations of commercial texts through their own media production work, examining how such production work involves critical analysis of media texts and use of texts as a way of defining and performing identities. Beavis discusses how educators might engage with and understand convergences between marketing and game companies. Beavis argues that educators need to know about the kinds of experiences young people are having in these spaces in order to help young people develop informed and critical perspectives on media industries. As marketing becomes increasingly individualised and covert, teachers need to be aware of how this aspect of the media landscape is changing for young people, what skills and knowledge young people are already developing, and what they need further to negotiate these environments. Clearly there is an increasing need for media literacy skills to be addressed in classrooms, as new challenges emerge in relation to digital cultures, such as issues around credibility, trust and ethics.

Marsh also highlights the need for schools to recognise young people's media literacy skills, using the metaphor of tectonic plates to describe ways in which educational environments might be creating greater gaps as the 'plates' of home and school knowledge connected with digital literacy practices move further and further apart. Marsh urges educators to engage in 'productive pedagogies' which draw on pupils' skills and knowledge, and she outlines key features of classroom work which exemplify the kind of connectedness with children's cultures which make classroom learning both meaningful and relevant. Carrington adds to Marsh's argument, suggesting that engagement with digital and traditional texts and technologies is becoming part of a hidden curriculum linked to longer-term academic success. There are issues here of equity and social justice, echoing those raised by Marsh, since if schools do not ensure that all pupils have such opportunities with a wide range of digital texts, then existing gaps in society both in terms of class and globally may well widen. Carrington emphasises the need for literacy educators to be aware of how the shifting values given to particular practices offer signposts to the ways in which literacy is being valued in the world beyond school. Mackey also suggests that educators would benefit by examining out-of-school literacy practices to see the multiple ways of engaging with texts and the different knowledge and understanding that develops through these engagements. Mackey's analysis of 'thick play', which characterises many of the practices discussed in this volume, suggests that young people are engaging in learning experiences which are different from those found in schools, but which schools might recognise and encourage.

Returning to Jenkins et al.'s (2007) report, we have evidence in this volume, as well as in research reported by Jenkins et al., that young people are developing new literacy skills in digital environments, and educational spaces have an important role to play in further developing media literacies. Jenkins et al. outline three concerns in relation to participatory media cultures which point to a need for educational intervention: the participation gap (unequal access to skills and knowledge); the transparency problem (learning to view media critically); and the ethics challenge (consideration of emerging ethical issues). As discussed above, the chapters in this volume provide examples of ways these concerns might be addressed in schools or further evidence of the need for schools to pay attention to these issues.

In the years since we started conducting research, we have seen tremendous changes in children and young people's engagements with media cultures, and with the advent of digital technologies, changes are occurring at a rapid pace. Change can highlight many interesting questions, and it is crucial that researchers and educators continue to identify and investigate issues which will help us to understand children and young people's engagements with digital technologies as well as to identify when and where further changes and adjustments need to be made. Research is showing that issues of access are complex, and as Jenkins et al. (2007) highlight, the participation gap needs to be recognised and addressed by educators and policy makers. Pedagogical issues are raised as new kinds of learning happen outside school, and new concepts become relevant and urgent for young people to understand. We hope that this volume will encourage others to build on the work included here so that we continue to gain new insights into this fast-developing and significant field.

REFERENCES

Jenkins, H., Clinton, K., Purushotma, R., Robison, A. J., & Weigel, M. (2007). *Confronting the challenges of participatory culture: Media education for the 21st century.* Retrieved November 27, 2007, from The MacArthur Foundation Website: http://www.digitallearning.macfound.org/atf/cf/%7B7E45C7E0-A3E0–4B8 9-AC9C-E807E1B0AE4E%7D/JENKINS_WHITE_PAPER.PDF

Millar, S. (1968). *The psychology of play.* Harmondsworth: Penguin.

Robinson, M. (1995). *Children reading print and television.* London: Falmer Press.

Sutton-Smith, B. (1997). *The ambiguity of play.* London: Harvard University Press.

Wade, A. (2007). *Blurred boundaries.* Retrieved January 12, 2008, from Media Guardian Website: http://media.guardian.co.uk/digitallaw/story/0,,2225125,00. html

Willett, R. (2005). 'Baddies' in the classroom: Media education and narrative writing. *Literacy, 39*(3), 142–148.

Winn, M. (1985). *The plug-in drug* (2nd ed.). Harmondsworth: Viking Penguin.

Winnicott, D. W. (1971). *Playing and reality.* London: Tavistock Publications.

List of Contributors

Shakuntala Banaji is a researcher and lecturer at the Institute of Education, University of London. She is the lead researcher on the European Union funded project 'Civicweb: Young People, Civic Participation and the Internet'; she has previously worked on a review of literature outlining Rhetorics of Creativity in writing about the arts and education, evaluating a set of media education materials about advertising aimed with primary school students called Be Adwise2 and as a consultant researcher on the project Children in Communication about Migration (CHICAM). She taught English and media studies to secondary school students in South East London for 6 years and has guest lectured on Hindi Film, taught on the media and society BSc at South Bank University and now teaches on the MA in media culture and communication at the Institute of Education.

Catherine Beavis is an associate professor in the Faculty of Education, Deakin University. Her research interests focus on literacy, popular culture and ICT; the changing nature of literacy and text as reflected in online popular culture; and the implications of young people's engagement with digital culture for curriculum and pedagogy in schools. Recent and current research projects include studies of the use of computer games as text in the secondary English classroom, young people's literacy and communication practices as they engage with online multiplayer computer games, intersections between literacy, community and identity in online popular culture and web design, and an investigation into gendered dimensions of computer game play, in the classroom and in internet cafés.

Victoria Carrington is Research SA Chair in Centre for Literacy, Policy and Learning Cultures and the Hawke Institute for Sustainable Research at the University of South Australia. She writes extensively in the fields of sociology of literacy and education and has a particular interest in the impact of new digital media on literacy practices both in and out of school. Issues of access and power crosscut much of what she researches and writes. She is on the editorial boards of a range of journals and is an editor of the international journal *Discourse: Studies in the Cultural Politics of Education*. Recent publications include co-editing a special edition of *Discourse: Studies in the Cultural Politics of Education* on digital literacies with Dr. Jackie Marsh (2005); 'Txting: The end of civilization (again),' *Cambridge Journal of Education*, Vol. 35(2), Summer 2005; 'The uncannny, digital texts and literacy' in *Language and Education* (2006); and the monograph 'Rethinking middle years: Early adolescents, schooling and digital culture' (2006).

Beth Cross is a research fellow at the Adam Smith Research Foundation, University of Glasgow, United Kingdom, with an interest in children's literacy issues, popular culture and the interface between formal and informal learning currently working on a three year ethnographic study of children's voice, participation and citizenship in the transition from primary to secondary schooling. Dr. Cross's work also includes consulting widely across Scotland developing multi-disciplinary and participatory approaches to storytelling with work with children.

Julia Davies is a senior lecturer in education at The University of Sheffield, where she directs the online MA in new literacies and the MA in educational research. Her research predominantly focuses on digital literacies; she is exploring uses of new technologies and the ways in which these are impacting on learning and literacy. Recent work has included a chapter about eBay and the digital literacies of online shoppers; the ways in which digital photography is impacting on the way we look at the world and our place within it; and academic blogging.

Clare Dowdall is a lecturer in education at the University of Plymouth where she teaches language and literacy modules. Currently she is engaged in doctoral research that explores pre-teenage children's digital text production and social identity work in online contexts. Her research interests pivot around the tensions that can be perceived between the formal curiculum for young children's education and their learning in online spaces. She has written about children's voluntary out-of-school text production, considering the features that may impact on this informal process. In addition, she has written about the tensions that can be observed between children's digital text production in school and non-school contexts.

Laurent Durup is project manager at OUAT Entertainment, a company specialised in trans-media products (video games, mobile games and applications, interactive DVD, DVD games, animated series, interactive TV). He holds a master's degree in management of children's products, delivered by The European Centre for Children's Products (University of

Poitiers). He had previously completed a bachelor's degree in history at Sorbonne University (University of Paris 1).

Margaret Mackey is a professor at the School of Library and Information Studies, University of Alberta. She researches, writes and teaches in the interdisciplinary field of young people's literacy and literature in both print and other media. Her research interests include: reading processes; interpretive approaches to print, graphic, digital and media texts; children's and young adult literature; popular culture and young people; and commodities and merchandising for youth. Her most recent book is *Mapping Recreational Literacies: Contemporary Adults at Play* (Lang, 2007).

Jackie Marsh is Professor of Education at the University of Sheffield, United Kingdom, where she directs the EdD. Jackie is involved in research which examines the role and nature of popular culture, media and new technologies in early childhood literacy, both in- and out-of-school contexts. She has conducted a number of studies that have explored children's out-of-school learning in relation to their use of media and new technologies, including the 'Digital Beginnings' study, a national survey of young children's use of media in England (http://www.digitalbeginnings.shef. ac.uk/). Publications include *Making Literacy Real: Theories and Practices for Learning and Teaching* (with Joanne Larson, Sage, 2005) and the edited volume *Literacy and Social Inclusion: Closing the Gap* (with Eve Bearne, Trentham, 2007).

Caroline Pelletier is a research fellow at the Institute of Education, University of London. Her work focuses on how new media technologies shape conceptions of knowledge, learning and teaching. She recently managed the project 'Making Games,' which developed approaches to teaching and learning about computer games in media and English classrooms and created production software to enable young people create their own computer games.

Muriel Robinson is principal and professor of digital literacies at Bishop Grosseteste University College, Lincoln. Her interest in this area grew out of PhD work looking at the ways in which children make sense of narratives in print and on television which suggested that there are many similar strategies being deployed (Robinson, 1997). More recently, she has worked with Margaret Mackey developing the idea of an asset model of literacy, namely a model which starts from the experiences, knowledge and skills that children have to draw on in any one situation rather than the more common deficit model which views popular culture as a negative influence on schooled literacy [with Mackey in N. Hall, J. Larson and J. Marsh (Eds., 2003) *Handbook of Early Childhood Literacy*;

with Turnbull in J. Marsh (Ed., 2005) *Popular Culture, New Media and Digital Literacy in Early Childhood*; with Mackey in J. Marsh and E. Millard (Eds., 2006) *Popular Literacies, Childhood and Schooling*]. She has also explored in the ways in which the next generation of primary teachers in England is being prepared and the extent to which they can be seen as Prensky's 'digital natives' (2001).

Valérie-Inés de la Ville is professor in organisation studies and business policy at the University of Poitiers, France. She is the founder of The European Centre for Children's Products—a training and research unit focused on children-orientated markets—that delivers a master's degree in management of children's products. She has edited in 2005 *L'enfant consommateur,* Vuibert, Paris, and coordinated as a guest editor a special issue on child and teen consumption in the *Society and Business Review* in 2007. Her research areas are in the collective foundation of entrepreneurial and strategic undertakings, the dialogical processes of strategy formation, and the strategic innovations in children-orientated markets as well as in the ethical issues raised by addressing children as consumers or economic actors within contemporary society. She currently supervises a national interdisciplinary research programme on children and fun foods.

Rebekah Willett is a lecturer at the Institute of Education, University of London, where she teaches on the MA in culture, language and communication. She is also a researcher at the Centre for the Study of Children, Youth and Media, which is based at the London Knowledge Lab. She has conducted various research projects on children's media cultures. Her research interests include gender, digital technologies, literacy and learning. She has published articles in journals, including *Discourse, Gender and Education; Journal of Media Practice; Learning Media and Technology;* and *Literacy;* and has chapters in several recent books. She also co-edited (with D. Buckingham) *Digital Generations: Children, Young People and New Media* (2006).

Index

A

Abbs, P. 208
ABC Family 41
Abusing Babyz.com 116
achieving global reach 36–53
acknowledging thick play 104–6
Active Worlds 202
AdAge China 19
adaptations 97–8
aesthetic hybridisation 24
aesthetic of unfinish 97, 138
affinity spaces 121, 167
agency 55, 125–42
AI *see* artificial intelligence
Al-Jazeera 59
Albright, J. 150
Alexa 74
Alvermann, D. E. 15, 200
amateur spoofs 54–67
ambiguity 64, 121
analysing principles of game design 175–9
anonymity 4
another world 129–33
anthropomorphic cakes 117–19
anticipation bandwagon 96
antisocial behaviour 87
Appadurai, A. J. 1
appeal of thick play 92–107: appeal of being the expert 99–101; building fictional commitment 94–8; implicit expertise 101–2; making a bigger world 103–4; selective passions 98–9; slippery texts 102–3; why it matters 104–6
archi-textuality 39, 46
Arendt, H. 140
articulations 7
artificial intelligence 153

Atomic Betty 37
Atwood, M. 73, 75, 79, 84, 87–8
Austen, Jane 99
authorship 39
autonomy 21
avatars 3, 21, 201

B

baby boomer generation 88
Babyz Community 109–111, 115–16, 122
background knowledge 204–7
Bahr, M. 9, 200, 203–4, 207, 216
Bakhtin, M. 39, 113, 125, 128, 137–40
Balshaw, M. 162
Banaji, S. 139, 147, 160, 221, 223
Barbie Girls 201–2
Barthes, R. 39, 184, 188
Barton, D. 75, 190, 197
Bateson, G. 64, 169
Batman 46, 94
battlefield of rhetorics 140
Bauman, Z. 75, 77, 80, 87
BBC 59, 74
BC *see* Babyz Community
Bearne, E. 76, 200
Beavis, C. 55, 225
Bebo 74–5, 80, 83, 205, 208
Beetlestone, F. 159
being the expert 99–101
Benkler, Y. 112, 120
Bentley Bros 61–3
Bentley, T. 152–3
Bernstein, B. 210
Bhabha, H. 64–5
Big Brother 59–60, 120
big worlds 92–107
binary opposition 149

Birmingham Centre for Contemporary
    Cultural Studies 150
Bixler, P. 97–8
Blair Witch Project 189
Blake, William 151
blank parody 57
blending spaces 112
Blizzard 18
blockbusters 16, 105
blogging 1–2, 55, 72, 110–117,
    120–21, 146, 150, 162, 186–7,
    192–5, 202–216
blogosphere 112
BMW 18
board game analysis course 170–71
Bober, M. 73–4, 205
Bobos 18
Boden, M. 150, 153
boiler room 59
Bortree, D. S.
Bourdieu, P. 78–80, 86, 101, 200, 202,
    216
boyd, d. 74, 112–13
bozbozboz 114–15, 117
brand cultures 17–18, 32
branding inter-textuality 50
breaking chains 115
Brennan, C. 155
Brewer, D. A. 103
bricolage 56, 216
British Empire 130
British stoicism 187–8
Brogan, C. 193–4
Bromley, H. 2
Brooks, G. 201, 205, 208
Bruner, J. S. 167
Bruns, A. 209
Buckingham, D. 2, 15–16, 18, 65,
    109–110, 126, 152, 158, 166,
    169–70
'Bug Busters' 63–6
building fictional commitment 94–8:
    adaptations 97–8; revisiting the
    same text 94–5; series reading
    95–6
Bulfin, S. 201
Bullen, J. 121, 128
bunnyhopping 59
Burn, A. 2, 139, 147, 158, 160, 166,
    170
Burnett, C. 205
Burns, E. 21
BurpsLiberty 115
Bush, G. W. 185

Butler, T. 200
Byron Review. 207

C
C-Monster 115
Caillois, R. 60, 166–8
Cain, C. 77–8
camcorder cultures 57–8
Campbell, J. 160
carnivalesque 138, 174, 178
Carr, D. 170, 226
Carrillo, R. 210
Carrington, V. 73, 76–7, 86, 88, 207,
    222, 224
Carruthers, P. 155
Carter, M. 76
Cartier-Bresson, H. 189
Cartoon Network 37–8, 43
Cartwright, L. 188–9
Castells, M. 77
categories of childhood 55
Cat's Eyes 41, 45
challenges of the future 219–27
Championship Manager 169
Chandler, D. 40, 56
changed communicational landscape
    15–16
Chant, D. 9, 200, 203–4, 207, 216
'chasing boys' 3
chatrooms 3–4, 6, 92
Chess 170–71
children and digital cultures 69–142:
    exciting yet safe 92–107; mime-
    sis and spatial economy of play
    125–42; online connections and
    collaborations 108–124; texts of
    me and texts of us 73–91
Christensen, C. 127
chronicles 108–124
Churchill, W. 141
Ciechanowski, K. 210
circuit of culture 7
Clark, C. 200
class-based stereotypes 59
Clements, D. H. 158
Clinton, K. 190–91, 193, 196–8, 225–7
clip culture 54
Club Penguin 201–2
Cluedo 170–71, 175–9
Coca-Cola 18–33
Code Lyoko 43–5
Cohen, G. 151
cohesion 110–113, 119, 121–2, 152
Coke *see* Coca-Cola

collaborations 108–124
collaborative narratives 110–111
Collazo, T. 210
collective intelligence 191, 195, 225
Collins, B. 186
Comber, B. 150
comic books 17
commodification 51
complexly coded cultural artefacts 188
computer games 2, 8, 13, 15–18, 25, 32, 93
Computer Mediated Communication 20
conceptions of play 166–9
conflict 121
connectedness to the world 204, 207–9
Consalvo, M. 22–4, 50
consequences of spatial economy 125–42
consumption 7, 54–67
contexts of digital cultures 11–68: achieving global reach on children's cultural markets 36–53; consumption, production and online identities 54–67; games within games 15–35
contextualisation 25–6
convergence 15–35, 55, 62: and media literacy 16–18
Convergence Culture Consortium, Massachusetts Institute of Technology 17–20, 22, 32
Cook, D. 55
coolness 24, 65
Cordes, C. 156–8
Craft, A. 150–51
craft of the classroom 158
creation of cool 19
creative affordances of technology 157
creative classroom 147, 149, 158, 161
creative industries 152
creative potentials of technology 148, 157–8
creative socialisation 152–3
creativity 125–42, 147–65: a digital 'creativity pill' 157–8; evaluation, learning and pedagogy 158–61; rhetorics of creativity 147–8; serious or playful stuff? 153–7; socialisation and 'successful' societies 152–3; unique or democratic? 148–51
creativity construction through discourse 147

Cremin, T. 150
critical literacy 15–35
Cronin, J. 186
Cropley, A. J. 150
Cross, B. 136, 138, 222–4
cross-fertilisation of ideas 115
cross-platform referencing 16, 18
crossings 108–124
Crystal, D. 118, 120–21
Csikszentmihalyi, M. 93, 150
cultural capital 79, 101, 132, 135, 200, 202, 216
'cultural odor' 51
cultural politics 162
Cunningham, H. 150

D
Daisuke, O. 2
Danath, J. 113
Davies, J. 76, 78, 109, 112–13, 119, 194, 203, 207, 222, 224
Davis, T. 212
Dawes, L. 167, 169
Dawkins, R. 114–15
day-in-the-life documentary 60
de Block, L. 65
de Certeau, M. 51, 56
de la Ville, V. I. 22, 24, 45, 225
De Mooij, M. 22
De Peuter, G. 45, 49–50
Dear, P. 183, 187
decisive moment 189
democratic creativity 148–51
Demos 136, 208
Department for Children, Schools and Families 207
Department for Education and Skills 76, 168
depthlessness 57
design objectives 45–8
developing affinities 112
DfES *see* Department for Education and Skills
dialogical activity 127
different galaxy in the back green 129–33
Digimon 44, 47
Digital Beginnings 201
digital 'creativity pill' 157–8
digital cultures in the classroom 200–208
digital cultures, play, creativity 183–99: image itself 188–9; images in context 184–8; implications

for literacy 192; participatory
culture 190–91; some signposts
192–7
digital divide 110
digital generation 109
digital literacies 111
Digital Research Youth Project 76
digital terrains 135–6
digitality 111
digitally mediated sociability 135
Disney group 41
distributed cognition 191, 195
distributed narrative 119
divergence 22
divergent thinking 155
Dixon, S. 155
DIY Internet 215
double voicing 113
Dovey, J. 64
Dowdall, C. 75, 112, 222, 225
drama-in-education movement 1
DrRob 120–21
'dry heaves' 99
Du Gay, P. 7
dual construction of fund raising 37–8
Dumbleton, T. 167, 169
Dupuis Audiovisuel 37
Durran, J. 166
Durup, L. 22, 24, 225
Dyer-Witheford, N. 45, 49–50
Dynasty Warriors 131
Dyson, A. H. 2, 213

E
easy rhetoric 109
eBay 115
Economic and Social Research Council 2
education, games and game play
166–82: board game analysis
course 170–71; conceptions of
play 166–9; conceptions of play
in analysing principles of game
design 175–9; 'Making Games'
Project 169–70; playing and
analysing Snakes & Ladders
171–5
Edwards, M. 160
Ehrmann, J. 168
Eldorado 36–7
elements of experience 154
Ellis, L. 210
Elwood, J. 65
email 46, 58, 74, 88, 108
embodied sociability 135

emotional investment 32
empowering model of creativity 6
enactments of play 166
encountering play and creativity in
everyday life 1–10
endorsements 21
Enlightenment 147–8
Epstein, D. 65
Epstein, R. 187, 288
erosion of civil rights 186
ethical symmetry 127
ethos stuff 111, 122
evaluation 158–61
everyday creativity 151
evidence for global reach of French
animation 36–7
exciting yet safe 92–107
exclusion 77
existing standards 38–40
experiential marketing 17
exploring the rhetorics and realities
147–65
extrinsic motivation 168

F
Fabos, B. 76
Facebook 87, 111–12, 116
Facer, K. 15
'fake memory of me' 116
fan cultures 17–18, 32
fan fiction 105
fan-service 45–7
Farley, R. 167
faux-Asian aesthetics 22–3
feminism 23, 28
femininity 65
Fernback, J. 121–2
fictional facts 100
Fiske, J. 60
flatness 57
flattening effect of classrooms 16
Fleckenstein, K. S. 102–3
Flickr.com 76, 110–111, 193–4
flirting 4
'flow' experiences 93
Forrester Consumer Technographics 21
Fox 41
France Animation 37
Freebody, P. 15, 191
French animation productions 36–40:
fund raising 37–8; inter-textuality 38–40; some evidence 36–7
French touch 22
Freud, S. 167

Friendster 74
fund raising 37–8
Funky Cops 37
fusion 23

G

Game Boy 102
gamers 75
games within games 15–35: conver-
    gence and media literacy 16–18;
    glocalisation and hybridity 22–4;
    World of Warcraft meets Coke
    18–21; *World of Warcraft* and
    year eight 25–32
Gardner, H. 153–4
Gauntlett, D. 150
Gee, J. P. 15, 77–8, 111, 167
Geertz, C. 93
Geist, M. 54
genealogy of narratives 187
Genette, G. 39, 48
Ghost busters 64
Giddens, A. 77
girl power ideologies 21, 33
Glebe School Project 155
global localisation 23
global reach on children's cultural
    markets 36–53: layers in inter-tex-
    tuality in French animation series
    40–45; mastering stakes of inter-
    textuality 45–51; recent French
    animation productions 36–40
glocalisation 22–4
Goldman, L. R. 127
Google 54, 57, 193
Grahame, J. 169
Grand Theft Auto 134
'Great Escape' 118
Green, H. 137
Groos, K. 167
Gulliver's Travels 103

H

habitus 79–80, 216
Hagood, M. C. 15, 200
Hall, C. 160
Hall, S. 7, 128
Halverson, R. 15
Hamilton, M. 75, 190, 197
Hannon, C. 137
hard-core masculinity 65
harmful play 162
Harry Potter 16, 62, 96, 116
Harvey, D. 77

healthy play 162
Healy, J. 2, 158
Heathcote, D. 1
Hebdige, D. 55
Hermans, H. 127, 197
Hey, V. 65
Highlander 98–9
Hills, M. 64
hip global identities 19
Hip Hop Hedz 18
Hobbs, R.
Holland, D. 77–8
Holloway, S. 126
Hopkins, S. 21
Howard, J. 185
Hughes, J. 201, 205, 208
Hughes, T. 125–6, 137–40
Huizinga, J. 60, 167–8
humanism 141
Huntingdon, S. 185
Hutcheon, L. 97
hybridity 22–4, 50
hyper-textuality 39, 47
hyperconnectivity 21
hypertext links 113–14

I

iconography 184–8
identity creation 80
Illinois Loop 149
image itself 188–9
images in context 184–8
imagination 137
imaginative divide in digital terrains
    135–6
implications of digitality for literacy
    192
implications of play for education
    166–9
implicit expertise 101–2
importance of imaginative terrain
    136–41
improvisation 73–91
Ince, M. 77
inclusion 77
individuated literacy 111
inner reality 1
insider status 23, 110
insider vocabulary/jargon 119–21
Instant Messenger 205
insults 4
inter-textuality 22, 24, 36–54, 118:
    playing with existing standards
    38–40

interior–exterior imaginative terrain
    136–41
intimacy 65
iPods 18
irreverent play 16
Ito, M. 2, 76, 186, 202
iTunes 21

J

James, B. 65
James Bond 16
James, L. 7
Jameson, F. 57
Janes, L. 7
je ne sais quoi 101
Jeffrey, G. 151, 159
Jenkins, H. 17, 49, 56–7, 61–3,
    190–91, 193, 196–8, 225–7
Jenkins, R. 78
Jewitt, C. 188–9
Johnson, A. 200
Jones, K. 152
Jurassic Park 213

K

Kant, I. 148–9
Katz, C. 128
Kehily, M. J. 65
Kelly, N. 102–3
Kennedy, H. 64, 128
Kenway, M. 121
Kirriemuir, J. 168
'kiss-chase' 3
Klein, N. 50
Kline, S. 45, 49–50
Klinger, B. 94
Knobel, M. 2, 111–12, 115, 122, 203,
    209, 216
knowledge economy 153
knowledge integration 209–210
Kogal 18
Koutsogiannis, D. 197
Kramer, K. 210
Kramer-Dahl, A. 150
Krashen, S. D. 105
Kress, G. 15, 73, 77, 79, 84–5, 88, 113,
    188, 190, 200
Kristeva, J. 38–9
Kwek, D. 150

L

Lachicotte, W. 77–8
lack of clarity 161
Ladwig, J. 9, 200, 203–4, 207, 216

Land, R. 15
Landry, C. 153
Langer, B. 51
Lankshear, C. 2, 111–12, 115, 122,
    203, 209, 216
layers of inter-textuality in French
    animation 40–45: *Code Lyoko*
    43–5; methodology 40–41;
    Totally Spies 41–3
Leander, K. 76–7
learning 158–61
learning ICT skills 7, 153
Leiken, R. 185
lettered representation 77
leverage 20
Lévi-Strauss, C. 56
Lewis, C. 76, 211
Lindstrom, M. 20
Lingard, B. 9, 200, 203–4, 206, 216
liquid life 87
Lister, M. 189
literacy 1, 15–35
literacy pedagogies 204–211: back-
    ground knowledge 205–7; con-
    nectedness to the world 207–9;
    knowledge integration 209–210;
    problem-based curriculum
    210–211
Little Women 98
Livingstone, D. W. 152
Livingstone, S. 1, 73–4, 205
ljc@flickr 115
Lord of the Rings 16, 99, 101
'Lost Childhoods' 208
Loveless, A. M. 157–8
ludus 60
Luke, A. 15, 191
Lunenfeld, P. 97
Lury, C. 55
Lyman, P. 76
Lyotard, J. 77

M

MacArthur Foundation 74, 76
McFarlane, A. 168
McGoldrick, C. 160
machinima spoof 59
MacKay, H. 7
Mackey, M. 2, 92, 94–6, 99, 101, 222,
    226
McKim, K. 74
McRobbie, A. 56
Madden, N. 19
magick 119

Maisuria, A. 150, 155, 159–60
making chains 115
'Making Games' Project 169–70
managing stakes of inter-textuality 36–53
Manford, P. 200
mantras 23
Marathon Animation 37, 41
Markham, A. 110
Marsh, J. 2, 15, 73, 75–7, 200–201, 205, 208, 223–4, 226
Marshall, B. 152
Marx, Karl 151
masculinity 65–6
'mash-up' 215
Massively Multiplayer Online Role Play Games 18–20
mastering stakes of inter-textuality 45–51: privileging kinds of textuality 45–8; towards integrated inter-textuality strategy 48–51
mateship culture 65
Matrix 44
Matsuda, M. 186
Maw, J. 65
Maynes, W. 99
MCEETYA *see* Ministerial Council for Education, Employment, Training and Youth Affairs
meanings 61
meanings of mimesis in children's play 127–8
media inter-textuality 49
media literacy 16–18
Media Participations 37
Meek, G. 1
memes 114–19
mental aptitude 149
mentoring 192
Merchant, G. 2, 112–13, 202–3, 207, 212
meta-textuality 39, 47
metanarrative 118
metatropic mimicry 128–9, 131, 137–8
MIGUZI block 43
Millar, S. 211, 223
Millard, E. 2, 15, 200
Miller, E. 156–8
Mills, M. 9, 200, 203–4, 207, 218
mimesis 125–42: a different galaxy in the back green 129–33; imaginative divide in digital terrain 135–6; interior–exterior imaginative terrain 136–41; meanings

of mimesis in children's play 127–8; mimicking the digital 125–7; Pokémon on the playground 133–5
mimicking the digital 125–7
Mindscape 109
Ministerial Council for Education, Employment, Training and Youth Affairs 15
MMORPGs *see* Massively Multiplayer Online Role Play Games
mobile phones 2, 45, 73–4, 186–9, 192, 205
moblogs 183
mockumentary 58
Moje, E. 210
monolithic things 78
Monopoly 170–71, 175–9
Monty Python's Flying Circus 59
Moon, J. S. 15, 200
Moonscoop 37, 48
Morely, D. 23
morphing 28–9, 79
Moss, P. 140
mother-tongue languages 150
motivation 167–8
Mottram, M. 200
MP3 21
MSN 26
MSN Messenger 73, 82, 108
multimodal text 25
multimodality 113
multiplicity 73, 93
multivoiced social subjects 76
MySpace 54, 58, 74–5, 80, 82–3, 86, 108, 112, 205, 208
myth 137

N
National School Board Association 201
natural creativity 148
naturalisation 24
Nayak, A. 65
Negus, K. 7, 151
neo-tribes 18
new literacies 111–12
New Literacy Studies paradigm 75, 111
New London Group 75, 77
new media age 85
Nielson/Net Ratings 109
Nixon, H. 16
'no means no' 23
non-goal orientated play activity 154
North, S. 201

notions of creativity 220–23

O
Ofcom 205
Office for National Statistics 74
Okabe, D. 186
Oliver, M. 160
Once Upon a Time 5–6
online connections 108–124: blending
    spaces 11; collaborative narra-
    tives 110–111; hypertext links
    113–14; insider vocabulary/
    jargon 119–21; memes 114–19;
    multimodality 113; new lit-
    eracies 111–12; playful spaces
    109–110
online digital texts 73–91, 93
online identities 7, 54–67: Bentley
    Bros 61–3; 'Bug Busters' 63–6;
    camcorder cultures 57–8; survey-
    ing online spoofs 58–61; young
    people, identity and consumer
    culture 55–7
O'Reilly, T. 122
Otaku 18
out-of-school learning and digitality
    224–7
outer reality 1
Oxford University 87

P
Pahl, K. 77, 79
paidia 60
palimpsest 98
Palmer, S. 2, 208
pan-Asian aesthetics 22–4, 28
para-textuality 39, 46
parallel processes 31–2
Parker, D. 158
parody 47, 55–8, 60, 64, 66
participatory culture 16–17, 32, 57,
    190–91
participatory literacy 111
pastiche 57–8
Pattie, C. 205
pedagogy 158–61
Pelletier, C. 223
Petrie, P. 140
phantasmagoria 138, 168
Piaget, J. 167
Pickering, M. 151
piracy 62
play and creativity in classroom blog-
    ging 211–16

play, creativity and digital learning
    143–218: creativity 147–65;
    digital cultures, play, creativity
    183–99; education and games
    166–82; productive pedagogies
    200–218
play theory 60, 148
playful pedagogies 16, 18
playful spaces 109–110
playful stuff 153–7
playfulness 168, 194, 215–16, 224
playground *Pokémon* 133–5
playing and analysing Snakes & Lad-
    ders 171–5
playing with existing standards 38–40
poaching 56
podcasts 193, 203
Pokémon 38, 42, 47, 50, 102–3,
    130–35
polished performance 73–91
polysemic identity 108, 113
Pope, R. 153
porous leadership 167
porous texts 2
possibility thinking 150
postmodernism 57
potential convergences 8
Power Rangers 43
practitioner interpretations of play
    136–41
Prensky, M. 88, 110
pretending 4
Primary Framework for Literacy and
    Mathematics 76
principles of game design 175–9
privileging kinds of inter-textuality
    45–8
problem solving 167
problem-based curriculum 210–211
production 7, 54–67
production–consumption cycles 128
productive pedagogies 9, 200–218: play
    and creativity in classroom blog-
    ging 211–16; productive literacy
    pedagogies 204–211; tectonic
    plates of home and school
    200–204
produsage 209
Prout, A. 126–7
psychic health 64
psychological moratorium 167
psychology 139
public displays of connection 113–14
punctum 184

pursuit of cool 18–19
Purushotma, R. 190–91, 193, 196–8, 225–7

Q
Qualifications and Curriculum Agency 76
Quart, A. 56
Queensland School Reform Longitudinal Study 203

R
real world 18, 20–21, 33, 43–5, 51
recontextualisation 210
regulation 7
Reid, M. 158
reimportation of Asianised images 23
remix culture 209
representation 7
repurposing 2, 50, 62–3
'revealed tastes' 38
revisiting the same text 94–5, 104
rhetoric of cohesion 121
rhetoric of progress 5
rhetorics of creativity 8, 139, 167
Rifkin, J. 51
risk saturation 186
risk-taking 159
Ritchie, L. 201, 205, 208
Roberts, S. 201, 205, 208
Roberts-Young, D. 56
Robertson, R. 23
Robins, K. 23
Robinson, K. 152
Robinson, M. 223
Robison, A. 190–91, 193, 196–8, 225–7
Robson, G. 200
role play 1
Rowling, J. K. 116; *see also Harry Potter*
Rowsell, J. 77, 79
rubber stamp testing 160
Ruminatrix 118
Russ, S. 155
Russell, L. 160

S
safe chat 202
Salen, K. 170–71
Sarama, J. 158
Saville-Smith, K. 65
Scanlon, M. 158
scepticism 25
Schaffer, D. W. 15

Schoen, B. 96
Scholtz, A. 152
school practice and out-of-school learning 224–7
Schott, G. 170
science fiction 17
Scollon, R. 190
Scollon, S. 190
Scruton, R. 149
Second Life 219
Secret Garden 94–5, 97, 101–2, 104
Sefton-Green, J. 15, 158, 169
Seiter, E. 2
selective passions 98–9
self-defence 23
self-reliance 21
self-scrutiny 87
Seltzer, K. 152–3
semantic cohesion 112–13
semantics 113
semiotic theory 79, 85, 113
semiotics 188–9
sensitivity 127
September 11, 2001 185
series reading 95–6
serious stuff 153–7
sexism 27–8, 33, 115
sexualised behaviour 4
Shah, B. 160
Sharp, D. 213, 215
SHE 18–19, 21
shifts in affordances 85, 94
Shiga, J. 215
Shrek 46
Silverstone, R. 60
Simonton, D. K. 149
SimplyClare 120
Sipe, L. R. 2
skills cocktail 190
Skinner, D. 77–8
slippery texts 71, 103–4, 121
Smith, C. R. 2
Snakes & Ladders 170–76
social bonds 20–21, 178
social good 148
social languages 78
social networking 1, 5, 55, 73–91, 201, 208
social participation 16
sociodramatic play 109
Soep, E. 65–6
some signposts for the future 192–7
Sonic the Hedgehog 130
Sony 7, 23, 59

spatial economy across digital divides
      125–42
Spider-Man 46, 94
'Spin the Bottle' 170
spin-off products 51
spoofs 29, 54–67
Squire, K. R. 15
Star Wars 180
Starbucks 111
Starko, A. J. 159, 162
Statements of Learning for English 15
Steinkuehler, C. A. 15
Strauss, A. L. 204
Street, B. 111
structures 55
structuring dispositions 79–80
'stuff of our communication' 79
Sturken, M. 188–9
style cultures 17–18, 32
subjective experience 178
'successful' societies 152–3
Sugar Dudes 117–19
surveying online spoofs 58–61
Sutton-Smith, B. 1, 4, 121, 125, 128,
      133, 137–9, 141, 155–6, 167–8,
      174, 178–9, 223
symbolic resources 55

T
technological stuff 111
tectonic plates of home and school
      200–204
Tele Imaged Kids 37
text bricoleurs 216
texters 75
texts of me, texts of us 73–91
textual landscapes 86–7
textualities 2
The Simpsons 57–8
The Sims 169
*Their Space–Education for a Digital
      Generation* 136
thick play 92–107
third space 64–5
Thomas, A. 75
Thompson, C.
Thomson, P. 160
Thorne, B. 3, 76
Through the Viewfinder 115
Tobin, J. 51
Tolkien, J. R. R. 99, 101
Tom and Jerry 57
Totally Spies 37, 41–3

towards integrated inter-textual strategy
      48–51
Toxic Childhood 208
Toy Story 207
transmedia entertainment 17
transmedia inter-textuality 49
trapped underground.jpg 183–99
trashing 87
trauma 99, 139
TroisTetes 118
Turkle, S. 20
Tyner, K. 2

U
ubiquitous creativity 151
uncoolness 32
understanding play and engagement
      with digitality 223–4
UNICEF 208
unique creativity 148–51
Universal McCann 20–21
universality 197
US Department of Homeland Security
      185–6

V
Valentine, G. 126, 136, 205
van Leeuwen, T. 113, 183–4, 188–90
ventriloquilism 113
violence 66
virtual sit ins 21
virtual world 15, 20–21, 43–4
vocationalisation 160
VW 59
Vygotsky, L. 1, 127, 154

W
Wade, A. 219
Walker, J. 119
Walkerdine, V. 178
Walkman 7
Warner Bros 41, 62
Warry, M. 9, 200, 203–4, 207, 216
Web 2.0 122, 201–3, 209, 216
Weber, S. 155
Weigel, M. 190–91, 193, 196–8, 225–7
Wells, L. 189
Whiteman, N. 167
Wiccan community 119
Wikipedia 193
Wilkinson, J. 205
Willett, R. 56, 222–3
Williamson, B. 15

Willis, P. 150
Winn, M. 223
Winnicott, D. W. 1, 64–5, 223
Wolf, S. 150
Wolstencraft, H. 200
World of Warcraft 18–21, 25–32, 170:
    and Coke 18–21; and year eight
    25–32
writing cyberbodies 75
writings of the self 110–111

X
X-box 26

X.A.N.A. 43

Y
Yee, N. 20
young people and consumer culture
    55–7
youth subcultures 55
YouTube 54–67, 108
'YouTube' effect 54

Z
Zelda 130, 134
Zimmerman, E. 170–71